# Bread, Beer and the Seeds of Change

## Change

### Agriculture's Imprint on World History

# Bread, Beer and the Seeds of Change
## Agriculture's Imprint on World History

**Thomas R. Sinclair**
*Department of Crop Science, North Carolina State University
Raleigh, North Carolina, USA*

**Carol Janas Sinclair**
*Independent Researcher
Durham, North Carolina, USA*

www.cabi.org

**CABI is a trading name of CAB International**

CABI Head Office
Nosworthy Way
Wallingford
Oxfordshire OX10 8DE
UK

CABI North American Office
875 Massachusetts Avenue
7th Floor
Cambridge, MA 02139
USA

Tel: +44 (0)1491 832111
Fax: +44 (0)1491 833508
E-mail: cabi@cabi.org
Website: www.cabi.org

Tel: +1 617 395 4056
Fax: +1 617 354 6875
E-mail: cabi-nao@cabi.org

A catalogue record for this book is available from the British Library, London, UK.

**Library of Congress Cataloging-in-Publication Data**

Sinclair, Thomas R., 1944–
Bread, beer and the seeds of change :  agriculture's impact on world history
/ Thomas R. Sinclair, Carol Janas Sinclair.
    p. cm.
  Includes bibliographical references.
  ISBN 978-1-84593-704-1 (alk. paper) -- ISBN 978-1-84593-705-8 (alk.
paper) 1. Agriculture–History. 2.  Food–History. 3.  Crop science–History.
I. Sinclair, Carol Janas. II. Title.
S419.S47 2010
630.9–dc22

                                        2010003655

ISBN:  978 1 84593 704 1 (Paperback)
ISBN:  978 1 84593 705 8 (Hardback)

Commissioning editor: Rachel Cutts
Production editor: Shankari Wilford

Printed and bound in the UK from copy supplied by the authors by CPI
Antony Rowe, Chippenham

# Contents

# About the Authors

Dr. Sinclair is an international leader in Crop Science who has undertaken scientific research with cooperators on all continents. He received his BS and MS from Purdue University, his PhD from Cornell University, and an honorary doctorate from the University of Padua, Italy. He is currently on the faculty of North Carolina State University. His research has covered a wide range of cropping issues including drought, fertility limitations, climate change, and biofuels. Further, he has studied all the major crops of the world. With more than 40 years of research experience he has developed a unique perspective that allows him to understand the challenges of growing crops and to apply this information in a historical context.

Ms. Sinclair has a lifetime interest in foods and nutrition. She attended Purdue University and received her BA degree in English Literature from Ithaca College. She has lived and traveled in many countries of the world exploring local methods in food preparation and the cuisine of the countries. She has been involved in organizing menus to facilitate new experiences of a range of foods and food preparation. Exploration of historical approaches to food was a natural extension of this interest.

Tom and Carol have been married more than 42 years and currently live in Durham, NC, near their young grandson Noah Thomas Fine.

# Acknowledgements

We appreciate the discussions and direct input to this book of our family: Davy Sinclair, Andy Sinclair, Janas Sinclair, and Andy Fine. Tricia and Fred Gregory provided the inspiration for a wife-and-husband team to attempt a collaborative effort on a book. We also appreciate their review of the completed manuscript. Jerry Bennett, University of Florida, and Larry Purcell, University of Arkansas, offered valuable suggestions in their reviews of the manuscript. We appreciate the encouragement and support N. Michele Holbrook, Harvard University, gave us from the initiation of this book and her useful suggestions for its content.

# 1

# Introduction

*We must begin by stating the first premise of all human existence and, therefore, history, the premise, namely, that man must be in a position to live in order to be able to "make history". But life involves before everything else eating and drinking.*

Karl Marx
*The German Ideology*

*Undoubtedly the desire for food has been, and still is, one of the main causes of great political events.*

Bertrand Russell

"I eat what I eat", Carol's Polish mother would respond when we urged her to try any new food. Her diet focused on meat and potatoes, and no dinner was really complete without them. Her statement, however, carried profound insight. The foods eaten early in our lives are the foods desired the rest of our lives. Perhaps it is a trick of evolution that we are "programmed" to be comforted by the foods that successfully nourished us through childhood. Individuals with adventurous DNA – who might readily explore new food sources – may well have died when they sampled too widely and encountered one of the many plants and plant tissues filled with protective poisons.

Indeed, a major component of "cultural identity" is the staple foods that a society consumes. People have proudly identified themselves as members of a group defined by the starchy staple of their diet: potato-eaters of Northern Europe, maize-eaters of the Americas, rice-eaters of Asia. We eat what we eat! From the early days of society, the seasonal cycles of crop growth determined the rhythm of life with rituals and ceremonies at sowing, harvest, and during the dark times of midwinter when food became scarce.

The need of societies for their staple food has been a powerful force through much of history. Until very recent times, the absolute focus of life for nearly all people was to produce food. We are now living in a time of historic aberration. For the first time, obesity is a much larger health problem in developed countries than lack of food. Today's fortunate people never need to consider the option that supermarket shelves will not be fully stocked with virtually any food desired. Before this very recent era, the coming harvest literally determined "feast or famine". The concern for nearly all people was not "Am I getting fat?" but "Will the harvest be large enough to hold off starvation for one more year?"

The basis for political power in the past was claimed ability within the natural world to tip the balance towards "feast". Rulers often took on the status of gods themselves or at least claimed to have intimate contact with the gods. Good yields meant that the rulers had done their job and ensured political stability. When the pharaohs "caused" the Nile to flood or the Mayan chiefs "delivered" the rain, their power was entrenched. Bad weather, disease, or insect infestations resulting in poor yields were explained away with cultural myths.

Until recently, most societies could only afford the luxury of releasing just a very few people from the drudgery of food production. Yet the usual study of history focuses on the lives of only these elite rulers. The lives of the masses of people who struggled for survival are rarely considered. However, it was the lives of these masses – and what they ate – that were the motivating ideas of this book. When visiting monuments of past civilizations we found ourselves wondering about the lives of the vast majority of people serving the powerful, rather than the powerful themselves.

Both of us grew up in the agricultural environment of rural Indiana, and we were aware of the challenges in modern food production. We saw the work of growing crops assisted by machines and chemicals, but we wondered, "How was the food grown in past societies?" "What work was required by the farmers in antiquity?" and "What foods did they eat?" One of the motivations in undertaking this book was to better understand the food production in these past societies and how this top-priority activity influenced the course of history. The interesting history for us, and we hope the reader also, is not limited to political and military conflicts, but rather how societies sustained themselves with the essentials of food production.

Of course, what became very obvious after only a little research is that the lives of most people through history were focused solely on the necessity of producing food. The labor demands in growing crops and turning the grains into food and beverage were onerous and unceasing. All hands – men, women, and children – were needed for food production. Until very recently, all or nearly all the energy required in crop production and food processing was provided by human muscle. Everyone was needed in the fields for sowing, weeding, harvesting, and threshing.

Not surprisingly, the drudgery and pain of growing crops fell to the masses of people who were by definition the lower classes. These classes were labeled as serfs or peasants, but in reality these people were slaves to the system that required their physical energy to produce crops and food. There was no future for them except the continuous labor required to produce the current crop, and then the next one. Indeed, the labor requirements were so great in some systems that a major component of the work force was actually slave labor. The only future for the slave was pain and death, resulting from the work he or she was forced to do.

Only in the last 300 years has human labor been replaced to a significant extent first by animals, and then in the last 60 to 80 years by fossil fuel energy. In industrial agriculture, the replacement of muscle by machines has, of course, had a profound impact on human life. Instead of virtually everyone being required to produce food, now in the United States growing food is the main occupation of less than 2% of the population.

This productive 2% offers modern day consumers a variety of foods that was unimaginable to earlier societies. The monotony of people's lives in past societies was not relieved by variety in the foods they ate. As discussed in Chapter 4, the hard physical labor of growing crops and producing food required that these people consume roughly twice the calories we need in our current, sedentary life style. Their large caloric requirements had to be met by foods that could be readily grown and easily digested. The need to consume large amounts of calories focused the diet on grain-based foods that provided the calories: mainly bread and beer. Diets of people in most societies were commonly centered on one or two foods, which in nearly all cases were derived from starches from grains. One clear advantage of these grains was that they were readily fermentable. We discovered that until modern times fermented grain, i.e. beer, was likely a major source of calories for men, women, and children. The attractiveness of beer was no doubt enhanced by the mood-altering benefits of the alcohol in helping people to deal with a life of monotony. While the plant species differed among societies, the fundamental nature of the diet was amazingly constant until modern times. Not surprisingly, there were major health consequences from lack of variety in diet.

In researching this book, we identified a historic time line in technological progress in food production: The Seeds of Change. Enhancing technology through the centuries allowed either expanded crop production on new lands or increased yield on existing lands. While the challenges faced by each society were unique, there were common problems in food production faced by all societies. There has always been the threat of environmental degradation resulting from growing crops. Some of the environmental problems were self-inflicted; others were results of natural perturbations. Reviewing these past societies serves as a healthy reminder that crop production is always vulnerable to environmental vagaries, and this truth has not diminished in modern times.

To examine the topics introduced above, this book is organized into three parts. The first part (Chapters 2 through 6) examines background information, which we identify as the "The Seeds", and discusses the basics of cropping and food production that had to be resolved by all societies. Chapter 2 in this section considers the fundamental question of "Why Agriculture?" While agriculture was fully engrained before the existence of the societies discussed in this book, it is useful to understand the motivations that caused people to begin growing crops. Agriculture was an extremely demanding life, in contrast to hunting and gathering. How did it come about that people ended up growing crops? Since agriculture arose independently in a number of locations, it seems there must have been a universally compelling reason to take up agriculture.

Chapter 3 in Part I considers which plant species were selected to be the basis for food production in each society. In the societies we consider, the staple grain food was derived from seeds of a very limited number of plant species, all from the grass family. Staple species used by each society depended on the environment in which the crop had to be grown, and consequently these cropping systems often had major influences on the historical path of each society. The preparation of food as bread and beer from the grains of grasses is considered in Chapter 4. Chapter 5 examines the nutritional consequences to humans of basing a diet on a single grain crop. Finally, Chapter 6 considers the basic requirements in using grass species as grain crops: water, nitrogen, and weed suppression.

Part II is an examination of food production during the Golden Age of five ancient societies that evolved more-or-less independently, each with unique agricultural practices. Each society was influenced by the grass species available for food production, the climate in which the crops had to be grown, and the geographical constraints placed on their crop production. Societies such as the Sumerians (Chapter 7), Egyptians (Chapter 8), Chinese (Chapter 9), and Maya (Chapter 11) developed food production systems based on only one or two plant species each. The production of these crops had to be closely matched to the constraints of growing crops in their local environments. Powerful rulers rose to organize and control these societies so that food production was sufficient for a large, concentrated population. Geographical constraints of crop production tied the population to a geographical area. The contrasting case was the Bantu of Africa (Chapter 10). The Bantu had available plant species that acclimated to a wide range of environments, and they had no need to create a strong political and military empire to sustain crop production in a constrained area.

Part III examines a succession of western societies which reflected a progression in increasing technology. We consider the Golden Ages of five societies in this sequence: Athenian and Roman Empires (Chapter 12), feudal Europeans (Chapter 13), British (Chapter 14), and Americans since 1950 (Chapter 16). Chapter 15 discusses the key scientific and technological developments that led to the revolutionary changes in crop production and food consumption in the American society. The ability to either control large areas of crop land or increase production on their existing land led to a rise to power of these societies. Local geographic constraints were less of a concern than in earlier societies because technological developments allowed water and land transport of grains over great distances. However, protecting the environments to produce food and/or defending the transportation network to import grains were critical in sustaining these societies.

The American experience since 1950 is unique in the history of crop and food production. Throughout nearly all of history, food production and preparation had been completely dependent on human labor. There were selective substitutions of some human labor by animals in previous western societies, but muscle remained the basis for food production. Scientific and technological developments in the era from 1850 to 1950 (Chapter 15) resulted in a total revolution of crop and food production using fossil fuels. The consequences of the scientific revolution are examined in Chapter 16. Tractors tilled and sowed the land, large harvesters crossed the fields to gather the crop, and grain was shipped on trucks, trains, and barges to large factories manufacturing food for an urban society. Fossil fuels could be readily used to overcome the challenges in crop production resulting from limited water, limited nitrogen, and competition from weeds. Water could be pumped onto dry lands, nitrogen fertilizer manufactured using natural gas, and chemicals synthesized from oil to kill specific weed plants. Crop yields per unit land area of agriculture rose to levels four, five, or more times than had ever before been produced in history. Simultaneously, the diversity of foods available to people exploded. Taking advantage of the luxury of fossil fuels, a few percent of the population were able to produce all food desired by the remainder of industrial societies. The resource of fossil fuels, unfortunately, is in finite supply.

# 2

# Why Agriculture?

*Then the Lord planted a garden in Eden, in the East, and there he put the man he had formed. He made all kinds of beautiful trees grow there and produce good fruit. In the middle of the garden stood the tree that gives life and that tree gives knowledge of what is good and what is bad. A stream flowed in Eden and watered the garden ... Then the Lord placed the man in the Garden of Eden to cultivate it and guard it. He told him, "You may eat the fruit of any tree in the garden, except the tree that gives knowledge of what is good and what is bad."*

*... The woman saw how beautiful the tree was and how good its fruit would be to eat, and she thought how wonderful it would be to become wise. So she took some of the fruit and ate it. Then she gave some to her husband, and he also ate it.*

*... Then the Lord said, "Because of what you have done, the ground will be under a curse. You will have to work hard all your life to make it produce enough food for you. It will produce weeds and thorns, and you will have to eat wild plants. You will have to work hard and sweat to make the soil produce anything."*

Genesis, Chapters 2 & 3

Is it possible that the Genesis story might actually give insight about the progression from the hunter-gatherer stage of human development to agriculture? Before cultivation of crops began some 10,000 years ago, humans had successfully fed themselves for tens of thousands of years by hunting wild animals and gathering plant roots, leaves, berries, and seeds. Aspects of the hunter-gatherer life style were very desirable – a Garden of Eden – as compared to the life of the agriculturalists who followed them. Today, the hunter-gatherer way of life is often portrayed as a harsh and precarious existence. This image is probably fostered to some extent by television showing extant hunter-gatherer people who are relegated to some of the harshest environments in the world. The Kalahari tribesman and the Eskimo are shown struggling in their extreme environments in an existence on the edge of starvation and annihilation. Those hunter-gatherers that still exist in our modern world have been pushed to the fringe environments, where indeed the climate is severe and survival is challenging. However, even in these environments the hunter-gatherers of today have sustained a rich culture without the burden of tilling the soil.

In fact, it is likely that in ancient times those hunter-gatherer groups living in richer environments had a reasonably comfortable life. Seasonal migration by hunter-gatherer tribes took them to the areas where they knew food was likely to be available, if not abundant. It has been shown that a fairly modest amount of work was required to harvest the food from the wild. For example, a number of years ago Jack Harlan, a crop scientist at the University of Illinois, collected seeds from a wild stand of einkorn wheat growing in the Middle East using only a stick and basket. In one hour he was able to collect one kilogram (2.2 pounds) of clean grain. There was no need to cultivate the crop when grain harvests as great as this could simply be collected for a small tribe. Similarly, a hunt once or twice a week in many environments is likely to have provided the amount of meat needed by the tribe. For example, Colin Tudge (*Neanderthals, Bandits and Farmers: How Agriculture Really Began*) reported that, even in the harsh environment of the Kalahari, tribesmen could satisfy their meat needs by hunting on the average only six hours a week.

In addition to the comparative ease of obtaining food by hunter-gatherers in contrast to agriculture, there were other advantages for hunting-gathering. Diets were likely to have been much more varied, and hence more nutritious. Hunter-gatherers would have had a mix of animal and vegetable foods from a wide range of sources through the progression of seasons. In contrast, agrarian diets were based on only one or two staple crops, augmented with few other foods, and their grain-intensive diet resulted in mineral and vitamin deficiencies. Even though protein was adequate for agriculturalists, specific amino acids were deficient because meat was rarely eaten by most people. Only the few in the elite class would have the luxury of regularly eating meat.

Hunter-gatherers also lived in smaller communities as compared to agrarian communities in which a large labor force was needed to tend the fields. A viable hunter-gatherer community was commonly an extended family of only about 50 people. As a result, the tribe was less vulnerable to disastrous food shortages. The small size of the hunter-gatherer societies was to some extent a consequence of the life style. Children of these societies were likely nursed for a longer time than in agrarian societies because mothers in hunter-gatherer tribes did not have to expend huge amounts of time and calories working the fields. Natural birth control resulted from hormone suppression in lactating women. Out of necessity, hunter-gatherer groups had to remain small so that they did not quickly exhaust the food resources in each new location. Any collapse of the available food supply would have readily decreased birth rate, and maybe increased death rate of the tribe. The population of the tribe was small and self-correcting by the nature of the hunter-gatherer food supply.

Finally, individual self-worth may have been greater in small hunter-gatherer societies where decision-making was probably more participatory. There were neither land areas to be managed nor hierarchy needed to coerce workers to the harsh labor of growing crops. No pressure existed for a stratification of society into leaders and workers. Decisions could be made communally and people could participate in gathering and hunting as their abilities allowed. Hunter-gatherer societies are likely to have provided much more agreeable life styles than those experienced by the vast majority of people in agrarian societies. Indeed, the relatively relaxed life style of hunter-gatherers often led explorers coming in contact with aboriginal peoples to the conclusion that these people were "lazy".

The puzzle is why agriculture? Why would people abandon "The Eden" of hunting-gathering to pursue the demanding life style to grow crops? It is clear that the transition did occur in a number of places around the globe following the end of the last Ice Age. The search for the motivation for adoption of cropping has not led to a single clear answer. One hypothesis is that humankind is naturally attracted to accumulating possessions, whether utensils for cooking or icons associated with religious practices. An increasing burden of possessions might have made it attractive to limit migration. Local food supplies would need to be supplemented with managed plants.

A classical hypothesis is based on a spiral of increasing food production and increasing population. Following the last Ice Age, climates worldwide became milder and plant life thrived. Wild grasses that produced seeds attractive for human consumption would have been among the species reacting rapidly to fill new niches in the environment. These wild stands of grasses were a ready food source for the hunter-gatherers. With abundant harvests, areas where these plants grew would be favored locales to which the hunter-gatherer tribes returned. With repeated good harvests, the tribes would have grown in population. Some of the seeds could have been unintentionally sown, or even intentionally sown, to enrich the next year's harvest. The idea of this population-based hypothesis is that the crops themselves enticed people into agriculture: once the spiral had been initiated and more food led to more people requiring even more food, the path to agriculture was inevitable and Eden was lost.

While the above two hypotheses are plausible and probably played important contributory roles, another motivating force may have been an important trigger stimulating the transition from hunting-gathering to agricultural production of staple grains – humankind's strong desire for mood-altering substances. Many hunter-gatherer tribes had a proclivity for harvesting mood-altering plants from the wild. Tribal rituals and celebrations often associated with the consumption of fungi, cannabis, or coca leaves. Of course, an important component of this consumptive pattern was alcohol as a result of fermentation of plant materials containing sugars such as fruits, roots, and seeds. Seeds of wild grasses containing large amounts of sugar and starch when simply soaked in water and allowed to stand for a day or two will produce low levels of alcohol.

The discovery that cereal seeds could be fermented opened a whole new option for alcohol and food. Grains could be readily stored for use throughout the year, in contrast to soft plant tissues that had only a short season of availability. The collection and storage of grains for fermentation at any time of the year would have been a reason to encourage the growth of selected grass species. Of course, the collection and storage of large quantities of grain for year-round use would have discouraged a migratory life style. The recognition of the inherent advantage of grains in alcohol fermentation may have been a key piece of the "forbidden knowledge" that pushed people from the Eden of hunting-gathering.

Indeed, there may have actually been a genetic nudge in the drive to consume alcohol, and eventually take on the task of growing fermentable grains in abundance. Those ape-like creatures that were the ancestors of humans and proved best suited for survival may well have carried the genetic ability to identify ripe, even over-ripe fruit. Prior to seed maturation, fruits containing developing seeds are protected from premature harvesting by a number of properties, including chemical compounds that have adverse consequences for the consumer. However, once seeds inside the fruit are mature and ready for dispersal, fruit ripens and sends out aromatic signals that it is ready to be consumed. Those ape-like creatures that could smell the initiation of fermentation in the soft fruit flesh would have a distinct advantage. Fruit in its mature state would have been more nutritious both in calories and in available vitamins and minerals. A taste for the ripened fruit, including the presence of alcohol in the fermenting fruit flesh, may well have been a contributory survival trait in human ancestral DNA.

An affinity for alcohol may have also been a key advantage as early humans gathered into larger societies. Liquid – i.e. water – is essential to sustain life. However, sanitary water supplies were an on-going challenge as populations became dense and stagnant. Early populations settled in river valleys where the soil was easily worked and water was available to irrigate crops if necessary. The increasing population of humans and animals would have quickly caused the river water to become unsanitary. Consuming the available water carried high risk of serious disease. Those populations that fermented grain, adding alcohol to the water, would have had an advantage in avoiding diseases. The heating of water when brewing grains and the alcohol produced in fermentation would have been critical in providing a relatively safe beverage to consume. No community had potable water until very recent times – men, women, and children consumed fermented gruel or low-alcohol beer.

Fermentation of grain also results in several health benefits that can be attributed to the yeast in the mixture. Growth of yeast releases protein into the water-grain mixture, and especially increases the content of the amino acid lysine, which is commonly in low concentration in grains relative to the needs of human nutrition. In addition, yeast growth synthesizes niacin that is needed to avoid the disease pellagra (Chapter 5).

While the answer to "why agriculture?" cannot be unequivocally resolved, it is clear that all the necessary components came together so that humans initiated agriculture in several places in the world. In these diverse locations, a few grass species were domesticated by each society, the grains were intentionally sown and harvested, and these grains were essential to early agriculturalists to bake bread and ferment beer. In this book, we step forward from the early stages of grain production in various places in the world to the Golden Age of several of these early societies. Topics discussed include the environmental conditions that existed to allow abundant crop production, the technology that was applied in growing the crops, and the methods used in baking and brewing during the Golden Age of each society. Finally, we consider the critical changes in grain production capability that contributed to the demise of these Golden Ages.

**Sources**

Crowe I (2004) *The hunter-gatherers.* In: The Cultural History of Plants, G Prance and M Nesbitt (ed). Routledge, NY.

Diamond J (1997) *Guns, Germs, and Steel: The Fates of Human Societies.* W.W. Norton, NY.

Dudley R (2004) Ethanol, fruit ripening, and the historical origins of human alcoholism in primate frugivory. *Integrative & Comparative Biology* 44:315-323.

Evans LT (1998) *Feeding the Ten Billion: Plants and Population Growth.* Cambridge University Press, Cambridge, UK.

Harlan JR (1992) *Crops & Man.* American Society of Agronomy, Madison, WI.

Standage T (2005) *A History of the World in 6 Glasses.* Walker & Co., NY.

Symons M (2000) *A History of Cooks and Cooking.* University of Illinois Press, Urbana and Chicago.

Tudge C (1998) *Neanderthals, Bandits and Farmers: How Agriculture Really Began.* Yale University Press, London and New Haven.

# 3

# What Crops to Grow?

*Humble is the grass in the field, yet it has noble relations. All the bread grains are grass – wheat and rye, barley, sorghum and rice; maize, the great staple of America; millet, oats and sugar cane. Other things have their season but the grass is of all seasons ... the common background on which the affairs of nature and man are conditioned and displayed.*

Liberty Hyde Bailey
Founder, College of Agriculture
Cornell University

There are literally hundreds of thousands of plant species in the wild that are potential food sources for humans. As discussed in the previous chapter, hunter-gatherer societies consumed parts of many plant species. Hunter-gatherers explored plants in their environment, sorting out which plants could be eaten and what part of the plant made a desirable food. The selection of various plant components for food was based on an intimate knowledge of the life cycles of these plants. Wild plants often synthesized chemicals that discourage consumption until the time it is to the plant's benefit to have organs harvested. For example, plants evolved mechanisms to make fruits unattractive for foraging animals until the seeds inside had matured. Only when fruits and seeds are mature and ready for distribution are the chemical defenses altered so that these organs are attractive for consumption by animals. In this way, plant survival is aided in the dispersal of the next generation of seeds. However, many upset stomachs, and even deaths, must have been the price paid for gaining human knowledge about which plant parts were good to eat and at what times of the year these various plant tissues could be consumed.

The plant species ultimately selected as the basis for agricultural societies both in antiquity and modern times became limited to only a very few. The major staple species can be identified by a quick look in the cupboards of today's kitchens: wheat, rice, and maize (corn). These three species alone contribute two-thirds of the global production of plant products for food. They are used to produce a range of foods such as breakfast cereals, bread, corn chips, and beer. In the case of maize, much of the crop is used as feed for animals to obtain meat, dairy products, and eggs. In addition to the three major staple species, another five, namely barley, oats, rye, sorghum, and millet, are among the plant species that account for 92% of the caloric intake by humans. These grass grains are consumed both directly and indirectly as feed for animals. In the United States, sorghum and millet are used to feed animals and sold as food to be put out for birds. All eight of these plant species were the basic staple food of one or more societies in antiquity. What were the factors that so dramatically narrowed the choice of plant species from the hundreds of thousands of possibilities to only the eight that remain the fundamental foods of most societies up to modern times? What were the qualities of these eight grains that led to their selection as the basic food for so many people? In this chapter we will explore these questions.

The eight staple species – wheat, rice, maize, barley, oats, rye, sorghum, and millet – are members of the grass family (*Gramineae* or *Poaceae*) of plants. The grass family includes approximately 10,000 species ranging from very short-stature plants used for golf course greens and modern lawns to sugarcane plants that may grow to four or five meters (12 to 15 feet) tall. A common feature of all grass species is their long, narrow leaves with parallel veins. Most grasses produce a stem (culm); the bottom section of each leaf, called the sheath, wraps itself around the culm to strengthen the plant (Fig. 3.1). The top portions of the leaves, which are the blades, are displayed to capture light. The growing area of the grass leaf blade, positioned at the intersection of the sheath and blade, allows the blade to withstand damage (from foraging animals or lawn mower) and still continue to grow. Several grass species, including those selected as crops, produce relatively large seed heads at the top of the stem. Harvesting is greatly facilitated when the grain head is readily accessible above the bulk of the plant. Seed heads positioned at about waist height of humans also eased the work of harvesting. It appears that in antiquity grass plants in the wild were readily harvested by bending the seed heads over a basket and shaking or beating the grains from the head.

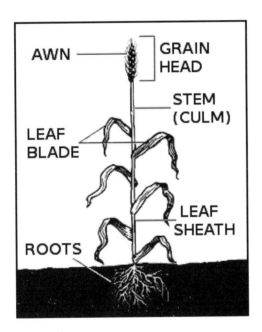

**Figure 3.1.** Diagram of the structure of a cereal-grain grass plant. This drawing is of a wheat plant; grass species vary in the number of leaves, shape and size of grain head, and presence or absence of awns. Maize plants differ in that the grain head has evolved to be a male tassel shedding pollen, and the female ears develop in the axils of leaves.

A side benefit of harvesting grass seeds would be the opportunity to store seeds for a long time since it was desired to have a product that could be used in fermentation year round. Storing bulk amounts of a plant product for many months almost invariably required that the plant material be low in water content. Even small amounts of water would have made the stored product highly vulnerable to attack by microbial pathogens. Fruits and vegetables can be dried for later consumption, but drying large quantities adds a great deal of work and increases the risk of spoilage in the preparation of the plant material for storage. Of course, some dried fruits were quite important in certain societies, for example dates in the Middle East, but these were invariably used as a supplement to the staple foods and usually consumed directly rather than fermented. Seeds did not require additional drying; the grain was immediately ready for storage upon harvest.

As suggested previously, for early agricultural societies an important aspect of harvesting and storing any plant organ was the capability for it to be readily digestible and easy to ferment. Hence, the material needed to be particularly high in carbohydrate content – sugars and starch. Many species in the grass family produce seeds of 85% or more carbohydrate content. The high carbohydrate content of the grass seeds also favored long-term storage. Seeds high in oil content, in contrast, can become rancid and undesirable for human consumption. For example, storage of peanut seed, which may be up to 50% oil, remains a challenge today. One area of modern research is to identify peanut varieties that have an oil composition that allows an extended "shelf life" for peanut foods. In antiquity those species that produced seeds very high in starch content, and therefore low in oil, would have been especially desirable.

Another advantage of grasses comes from their root structure, especially when water and nutrients in the soil are limited as was very often the case. Grasses have a root structure that readily spreads many roots both horizontally and vertically. These "fibrous" roots allow the plants to explore a large volume of soil, particularly in the upper soil layers, to accumulate water and nutrients. The fibrous root structure contrasts with many other species that have a tap root structure, which is characterized by a single, dominant root that penetrates deep into the soil.

**Box 3.1.** The Doomsday Seed Vault

Recently, seed storage has been taken to a new level to protect the valuable genetic diversity of crop species. These genetic resources have been collected both in the wild and from farmers who have passed seeds down through the generations. To store these valuable seeds, "seed banks" have been established around the world to store all seed types of each crop species. The genetic diversity among the thousands of lines that commonly exist for each crop species is critical in breeding projects to incorporate into commercial varieties such traits as desired growth habits, useful seed composition, and resistance to diseases and insects. However, a concern arose that concentrating all these valuable seeds in individual seed banks makes the collections vulnerable to catastrophic events such as fire, flood, or benign or terrorist human activities. As protection against such disasters, a new facility has been constructed to safely store backup seeds apart from all the seed banks. This backup facility was opened in February 2008 and officially identified as the Svalbard Global Seed Vault, sometimes called the Doomsday Vault.

The Global Seed Vault was constructed inside a mountain on the Norwegian island of Spitspergen in the Svalbard archipelago, which is a remote location without tectonic activity. The location within the mountain is at 120 meters (430 feet) altitude, which places it above any potential threat of global sea rise. Seeds are sealed into plastic packets to exclude moisture and stored at -18 °C (0 °F), which is a temperature that assures long-term storage of viable seed. The large vault, which has the capacity to store up to 4.5 million seed types, is kept sealed to protect against any outside dangers. Since the seed vault is located only 1300 kilometers (810 miles) from the North Pole, a failure in the refrigeration system will result in a temperature rise in the vault to only -3 °C (27 °F), still assuring a satisfactory temperature for long-term storage. The Global Seed Vault offers crucial backup protection for the crop genetic diversity needed to continue meeting not only the evolutionary challenges of diseases and insects but also the need to improve crop-yield capability in changing environments.

Fibrous roots rapidly and fully explore the top zones of the soil. If water and nutrients are predominantly available in the top layers of the soil, fibrous roots with their horizontal component of growth maximize the opportunity to recover these resources, even when plants are grown in wide spacing as was likely in early cropping systems.

Grass plants originally harvested by hunter-gatherers from the wild were already substantially changed when grown by the societies discussed in this book. Success in the wild by any plant species is dependent on survival of enough seeds to ensure a new population of plants in the next growing season. There are several key traits that aid in plant survival in the wild. One set of traits concerns seed dispersal. Grass heads in the wild "shatter" readily so that the seeds are quickly released once they are mature. This allows the seeds to fall to the soil and sow the next crop at that location. Further, some species developed characteristics that aid in seed dispersal to new locations by attaching seeds to animals' hair and fur. Some grass seeds are wrapped tightly in an outer hull that has an elongated extension called an awn, or beard. The awns can have tiny barbs that readily attach to the fur of animals. Consequently, when an animal comes to eat the mature seeds of the grass a few of the seeds might attach to the animal and be carried to new locations. Though useful in the wild, awns are a nuisance to human harvesters.

Another set of traits advantageous to the survival of wild grasses has to do with spreading the development of the next generation of plants over a span of time so the whole population of the next generation is not vulnerable to the same threat. Seeds mature on the mother plant over a long time period so there is a range of opportunities for seed dispersal. In addition, there are mechanisms in the seeds that cause varying time delays among the seeds before germination. Again, this trait enhances the possibility that at least some of the plants will successfully germinate and grow into the mature plants of the next generation. Both sets of these traits are undesirable to a grower who wants to harvest the crop on one or two passes through his field. For the domesticated crop, uniformity in both seed maturity and germination is favored.

Early farmers dating before the societies discussed in this book would have overcome these survival traits of the wild grasses to make them more useful in a cropping environment. Some selection is likely to have come about by the nature of the cropping situation. Those plants that had genes making them less prone to shattering would have remained intact and become an increasing portion of the harvested grain year after year. Ultimately, the shattering traits would have been minimized in the crops passed to succeeding generations of growers. Human harvesters also fostered the genetic trait of uniformity in maturity. The work of harvesting would be lessened if harvesting could be done on only one pass through the field. In addition to selecting for plants that had non-shattering heads, this harvest schedule would have also favored plants that matured their seeds at roughly the same time. These seeds would be harvested and some of them sown for the next year's crop. Any seeds that were past maturity or still immature when harvested could be eaten but would not germinate if sown to produce the next crop. Therefore, there would have been a gradual transition to selecting seeds for plants that matured their seeds in synchronization.

Other traits favored by human selection would have been ease in seed removal from the hull during threshing and production of large seeds. Grass seeds develop a hull that usually tightly encloses the seed. Since the hull is difficult to digest, those seeds that readily separated from the hull in threshing would have been favored and selected for the next generation of seeds. There also appears to have been a tendency to make selections resulting in smaller and less barbed awns. Large seeds were also likely to have been collected in the threshing process. Large seeds often result in more vigorous seedlings when the crop is sown.

The whole appearance of some species was changed with domestication. One of the most dramatic examples is maize. In its ancestral form, teosinte, maize had a grain head that was produced at the top of the stem like most grasses (Fig. 3.2). Teosinte plants grow as tall as, or taller than, humans; there was a mechanical weakness in these plants when selection for high yield produced a heavy grain head. Large weight on the top of these stems made the stem vulnerable to breakage when winds blew on the plants. Through a progression of mutations, the male and female parts of the maize became separated on the plant (Fig. 3.2). The relatively light-weight male part remained at the top of the stem as the tassel and produced the pollen that fertilized the female structures. The female part (i.e. ear) developed on these "new" plants in the axil of the leaves lower on the stem. One of the first maize mutations seemed to have resulted in a number of small grain ears in the leaf axils up and down the stem. Human selection, again probably in response to a desire to lessen the work of harvesting, favored plants that produced only two or three large ears within easy reach on the stem. The evolution of the maize plant continues in modern times through directed human breeding to produce large ears and plants that can be grown in crowded conditions at high plant populations. Like other grass species, many of the traits that enhanced survival in a natural environment have been removed such as a distribution over time in seed maturity and germination. These modern plants would have a difficult time surviving in natural ecosystems.

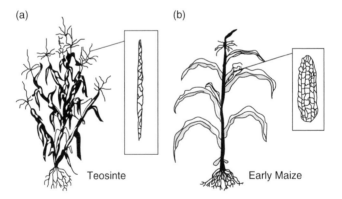

**Figure 3.2.** Drawings of ancestral teosinte and modern maize. The grain of teosinte is produced in the grain head at the top of the stem; on modern maize large grain ears develop in leaf axils. (From NativeTech (n.d.), courtesy of TaraPrindle/NativeTech.org)

It is clear that the crops that had been chosen in antiquity and ultimately used by the societies discussed in the book were not random selections. These plants were selected for specific traits that made them attractive for domestication and for production of grains that were desired for human consumption. Individual grass species had a universal appeal. One or two of the primary eight grass species provided the food base for nearly all powerful societies. These species had a number of desired traits with the most important being a seed high in carbohydrate that could be readily stored and used in making bread and beer. Human selection resulted in an evolution that left a legacy on which the societies discussed in this book could build their civilizations.

## Sources

Evans LT (1998) *Feeding the Ten Billion: Plants and Population Growth.* Cambridge University Press, Cambridge, UK.

Harlan JR (1992) *Crops & Man.* American Society of Agronomy, Madison, WI.

NativeTech (n.d.) *Native American History of Corn.* [Online] Available at: http://www.nativetech.org/cornhusk/cornhusk.html [Accessed 26 January 2010].

Perry GH, NJ Dominy, KG Claw et al. (2007) Diet and the evolution of human amylase gene copy number variation. *Nature Genetics* 39:1256-1260.

Prance G, M Nesbitt (2005) *The Cultural History of Plants.* Routledge, New York.

Solbrig OT, DJ Solbrig (1994) *So Shall You Reap: Farming and Crops in Human Affairs.* Island Press, Washington, DC.

Toussaint-Samat M (1987) *History of Food.* Translation by A Bell. Wiley-Blackwell, Oxford, UK.

# 4

# Basics of Beer and Bread

*Beer is living proof that God loves us and wants us to be happy.*

Benjamin Franklin

*And a special means of ye healthfulenesse of ye passengers by ye blessing of God wee all conveyved to be much walking in ye open ayre and ye comfortable of our food: ... we had no want of good wholesome beere and bread.*

Richard Mather, 1635
Puritan Minister

Growing crops in the field, harvesting and threshing the grain, and preparing the grain for consumption all require large amounts of energy. Until very recent times virtually all of this energy was supplied by human muscle. Therefore, those people directly involved in grain and food production – the vast majority of people in historic societies – had to be "fueled" by food and drink. This intense physical activity according to calculations in Chapter 5 would require consumption of 4000 to 5000 kilocalories daily. (The kilocalorie unit is shortened in the U.S. simply to calories in common use. To keep the units explicit, we will use kilocalorie as the energy unit. One kilocalorie is the amount of energy required to increase the temperature of 1000 grams of water from 14.5 °C to 15.5 °C.) Nearly all this energy was obtained by the consumption of these same grains. Mature grass seeds, however, are physically hard and when eaten directly would surely have taken their toll on teeth. Further, the unprocessed grains are not well digested. Therefore, in nearly all cases the grains were consumed after either soaking or grinding. When soaked in water the grain would become soft and could be eaten directly. If the soaked grains were allowed to stand for a longer time, fermentation would begin. In this book, we generically refer to the beverage derived from fermentation of these grains as "beer", even though modern beer drinkers would not recognize these as beer. Ground grain was used to make breads of all sorts.

The basics of processing grains to produce beer and bread were fully developed before the Golden Ages of the societies considered in this book. Indeed, the basic techniques in producing large quantities of beer and bread were prerequisite knowledge for structured societies to develop; people engaged in the hard labor of agriculture had to have a safe beverage to drink and a means to consume large quantities of calories. Each society selected specific procedures and which grain to use in producing beer and bread. There were some common basics, however, that had to be addressed in making these valuable products. In this chapter some of the basic steps for brewing and baking are presented in more-or-less a sequence of increasing complexity.

The hunter-gatherer's precursor to beer and bread in grain consumption was simple gruel or porridge made by soaking grains in water. Soaking the grains would soften them and also improve their digestibility. If during harvesting of the seed the grain had been cracked to some extent, this would aid in water infusion into the grain. In some species, in the early stages of domestication harvested grain hulls were tightly attached and were removed by pounding the grain with large sticks. Additional grain cracking may have been obtained by grinding the grain between stones. The cracked grain would then be soaked in water for one or two days to soften the grain.

Heating of the grain and water mixture would have produced a major improvement in the digestibility and nutrition gained from grass grains. Heating changes the structure of proteins in the grain; many of these proteins are associated with binding starch granules. Once the structures of the proteins are changed, i.e. denatured, starch granules can "unwind". This unwinding of the starch and gelatinization with heating exposes the molecules so they are much more accessible for digestion in the human intestines. Taste was also enhanced as sugar molecules would have been released into the gruel mixture.

Unintentionally, an additional benefit of heating gruel would be sanitizing the water used in the mixture. Temperature required to denature proteins in grain would also denature proteins in many disease microbes living in water. Depending on the temperature to which the gruel mixture was heated, the disease microbes would be substantially weakened or even killed. Those people who heated their gruel and avoided the direct consumption of water would have lessened their exposure to diseases carried in water. However, collection of fuel for heating placed another burden of work on these societies. Wood, plant residue, or animal droppings had to be collected, sometimes from considerable distances, to fuel heating of the gruel.

A very early discovery came as a result of allowing the gruel to sit for a day or so after it was heated. This additional time resulted in a major change in taste and impact on the consumer. Airborne single-celled yeasts (genus *Sacchromyces*) could settle on the exposed mixture of water and grain, and given warm temperatures the yeast would grow quite well in the mixture. Respiration of the grains and the yeast would rapidly consume oxygen in the gruel mixture, lowering its oxygen concentration. Oxygen concentration in the mixture remains very low because oxygen is slow to diffuse back into a liquid mixture from the air. Low oxygen concentration has the significant consequence of allowing yeast to digest sugar molecules by anaerobic respiration. The critical result of this reaction was that yeast released ethanol, i.e. alcohol, and carbon dioxide as by-products of anaerobic respiration. Of course, the discoverers had no concept of the microbial activity, even though this discovery became the essential biochemical process in preparing grains for beer and bread. The action of yeast is fairly fast, and if a gruel mixture is kept warm and allowed to stand for as little as 24 hours there could be a noticeable production of ethanol. Heated millet, for example, releases a fairly large amount of sugar, and the alcohol content might reach one or two percent in the first day. While this gruel would not be identified as a beer in modern times, it had several desirable qualities: consumed in large enough quantities the fermented gruel improved human nutrition, sanitized water, and probably most importantly provided a ready source of mood-altering alcohol.

Production of gruel-beer was a feature of a number of hunter-gatherer societies long before the societies discussed in this book. Indeed, as we argued earlier, the attractiveness of the mild mood-altering influences of gruel-beer was likely a strong motivating force for collecting grains from cereal species growing in the wild. At times of the year when these grains could be harvested, preparation of the gruel-beer was an important daily task. Once a gruel mixture particularly desirable in terms of speed of fermentation, taste, and alcohol content had been produced, the yeast could be retained for use in producing the next batch of beer. Ancient brewers certainly understood the advantage of carrying forward some of a desirable gruel mixture as an ingredient in making successive batches of beer, which they did by simply using a small portion of the original gruel with desirable yeast in the new batch of gruel. Continual reuse of clay pots with small cracks in their walls could also have been a means for transferring yeast between successive batches of gruel.

One of the outcomes of transferring portions of the beer product to succeeding batches of gruel was the possibility of isolating yeast strains that could produce increased alcohol content. While yeast produces alcohol, yeast cannot survive high alcohol concentrations in the gruel mixture. When the alcohol reaches the critical concentration for a specific yeast, it does not continue to grow. A batch of gruel with yeast that could produce alcohol concentrations greater than the basic one to two percent would be noticed, especially when drunk in quantity! In addition, higher alcohol concentration in the mixture offered more sanitary protection than lower alcohol concentrations. While alcohol concentrations of four to six percent ethanol by themselves would not sterilize water, they would contribute to cleansing the water and help thwart the reestablishment of disease microorganisms once the beer was brewed. The "shelf life" of the mixture would thus be extended a few days. Those yeasts that resulted in a four to six percent alcohol concentration in these early beers would surely have been valued and carefully handled for use in the making of the next batch of beer.

Yeast captured for producing higher ethanol concentrations would also require higher levels of sugar in the gruel mixture to support greater fermentation. Much of the work of the early beer brewers was to prepare grains so that the amount of sugar for fermentation was maximized. The exact means by which the grains were prepared to obtain high sugar yield varied a great deal among societies, depending to a large extent on the grain species being fermented. Since the carbohydrate in the cereal grains is stored mainly as starch, it was highly desirable to work out techniques to cause the starch to break down into sugar. In grass seeds, the breakdown of starch happens naturally when seeds are germinated so that sugar is released in the seed to provide energy needed in the metabolism of the structures of the new seedlings. Sugars are also used in the germinating seed directly as building blocks for many of the organic components of the growing seedling. During early germination of the seeds, enzymes called amylases are released in the seeds to disassemble the starch for breakdown into sugars for the growing seedling. Once seeds absorb water, the release of amylases is triggered. A key to brewing beer of higher alcohol content is to stimulate the natural action of the amylases to break down starch to release large amounts of sugar for use by yeast in fermentation.

The simplest approach to exploit the activity of the amylases in the grain is to simply allow the grain to begin germination. This process, malting, prepares the grain for fermentation. Malting is sometimes done with sorghum, millet, and maize by allowing a full seedling to develop from the seeds. The Bantu in Africa using sorghum and millet, and the Maya in Central America using maize, apparently malted these grains by allowing full germination of the seeds. Seeds were soaked in water for half a day or so and then spread out to allow germination. The germinating seeds were kept wet, and after about one week the starch in the grain would be broken down. The sprouted grains were dried and pounded into a meal to use in brewing. A limitation of this approach, however, is that the germinating seedling consumed a portion of the released sugars and the overall sugar yield was diminished.

One common approach to minimize sugar loss is to abbreviate the length of the germination process. Germination is allowed for only the short time required to release the amylases. Seeds are wetted, and after only a few hours the seeds are dried to stop germination. In the case of barley and wheat, seed germination is stopped at about the time that there is first visual growth of the root in the seed. The seeds are dried to obtain the "malt" for brewing. Green malt is the product obtained when the germinated seeds are simply dried. Roasted malt results when the germinating seeds are heated causing the malt to turn brown. The malt can be stored under dry conditions until needed for brewing. When ready for making the next batch of beer, the ground malt is mixed with water and often additional grain to increase the amount of starch in the mixture. A mixture of malted grain and additional grain allows the starch to be digested to release sugars.

Barley is especially popular in brewing beer because it has a unique combination of two types of amylases in its grain. Unlike other grains, barley has an abundance of both alpha-amylase and beta-amylase. The combination of these two enzymes is required to fully degrade starch. Not surprisingly, barley malt was mixed with other grains to enhance the sugar yield. Barley was identified as an important crop in several early agrarian societies.

Procedures for preparing mash, which is the mixture of malt, added grain, and water, varied a great deal among societies. It is clear, however, that heating the mixture would have greatly aided in the unwinding of the starch granules for attack by amylases. The Sumerians and Egyptians seem to have "half baked" loaves made from the malt and water mixture. The heating in this step could have provided the warm temperatures for gelatinization of the starch without reaching temperatures that destroyed the amylases. Later societies, which had developed clay pots that could be placed on fires, probably simply heated the mash. Assuming the temperature of the mash was checked with a bare hand, then the temperature of the mixture could have been kept around 50 °C (122 °F), which is moderate enough to avoid destruction of the amylases.

The fundamental step in brewing beer is to produce a mash that contains high sugar levels for yeast growth. Taste and appearance of beer were quite varied among societies as a result of differences in choice of grain, malting procedure, the specific yeast, and various flavorings added to the mixture. Temperature regulation during fermentation was another important variant. The addition of hops, which give many modern beers a distinctive flavor, is a fairly recent development (approximately 1000 AD) in beer making that is discussed in Chapter 13.

As in the case of beer, each of the societies discussed in this book had specific ways to bake bread from the grains they grew. The simplest technique for making breads was well established before these influential societies came into existence. These techniques involved grinding grain and mixing it with water to form a "dough". The grinding of the grain added greatly to the work load in producing food. Until recently, the grain was almost always ground with stone. An early approach was to use a quern that consisted of a saddle-shape rock on which the grain was placed and a grinding rock which was moved back and forth to crush the grain (Figure 4.1). A negative consequence of using rocks to grind was the rock dust that ended up in the crushed grain. Rock dust and particles wore away teeth after years of consuming the bread. The ground grain was mixed

**Figure 4.1.** A saddle quern consists of a top rubbing stone that was pushed and pulled over the grain placed on the base stone. (From Janick 2002, courtesy of Jules Janick.)

with water causing the starch granules to unwind, much as they did in the gruel mixture. If the dough was allowed to sit in a warm environment, gelatinization of the starch would be enhanced. Once the dough had reached a satisfactory texture, it could be flattened into thin cakes and baked on a hot surface or in an oven. Such flat breads were the staple in many societies, especially those that did not grow wheat.

Also paralleling beer production, the next step in increasing complexity in baking bread is to introduce yeast into the bread-making sequence. Yeast again introduces fermentation into the mixture. The key product in yeast fermentation in dough is not ethanol, but rather carbon dioxide. Carbon dioxide, released from yeast fermentation and trapped in the dough, causes the dough to expand or "rise"; this results in leavened bread. The amount of sugar needed to generate the required carbon dioxide in leavened dough is much less than necessary in brewing beer. Hence, enough sugar is usually directly available from grinding of the wheat grain so special procedures to increase dough's sugar content are not essential.

The critical condition to obtain leavened dough is that gas movement in the dough must be limited. Oxygen movement into the dough must be restricted so that anaerobic conditions develop within the dough. In this anaerobic environment, yeast initiates fermentation resulting in the release of carbon dioxide into the dough. The carbon dioxide gas trapped in the dough causes the dough to rise. Baking the risen dough locks in the structure of the bread and denatures the proteins resulting in a product with light texture and "yeasty" flavor. Baking also drives off the ethanol. As a result, leavened bread is usually a readily digestible food of a highly desirable taste.

The required elastic characteristics unique to wheat dough are a result of the glutens in this grain. Only wheat has sufficiently high amounts of glutens to effectively result in an inhibition of gas movement in the dough. To obtain leavened breads from other grains, glutens can be introduced into the dough by mixing wheat flour with flour of another grain. For example, rye bread is obtained by mixing wheat and rye flour. The proportion of each grain dictates to some extent the "heaviness" of the bread. A major advantage of mixing wheat flour with flour from another grain is that it also extends the wheat supply to increase the amount of bread that can be baked. For example, in northern climates wheat grows poorly as compared to rye, but a mixture of wheat and rye flour maximizes the amount of leavened bread that can be baked from a limited wheat supply.

As in beer, a key component in the baking of leavened bread is the yeast. The efficiency of a yeast strain in generating carbon dioxide and the taste it confers to the bread are important attributes. Portions of the dough mixtures that resulted in especially desirable bread could be saved for use in preparing the next batch of dough. Given the appeal of beer in many of the early societies, it is not surprising that it appears the same yeast was used for both beer and bread. In fact, the same building was used to brew beer and bake bread. As described above, it appears in ancient Egypt "half-baked" bread dough may have been used to introduce yeast and enhance sugar availability in the mash for brewing beer. Conversely, it may have been that beer was used to introduce yeast into the bread dough. In Sumeria and Egypt, barley and wheat were likely grown together as a single crop. The mixture of wheat and barley would complement each other in beer and bread making. The amylases from barley facilitated sugar production in the grain mixture and glutens from wheat allowed leavening of the bread.

---

**Box 4.1.** George Washington's Beer Recipe

"Take a large Siffer [sifter] full of Bran Hops to your Taste. Boil these 3 hours then strain out 30 gallons into a Cooler put in 3 Gallons Molasses while the Beer is Scalding hot or rather draw the Molasses into the Cooler & St[r]ain the Beer on it while boiling Hot. Let this stand till it is little more than Blood warm then put in a quart of Yea[s]t if the Weather is very Cold cover it over with a Blank[et] & let it Work in the Cooler 24 hours then put it into the Cask – leave the Bung open till it is almost don[e] Working – Bottle it that day Week it was Brewed."

---

The overwhelming source of calories for most humans until very recent times was a rather monotonous diet of beer and bread. The specific nature of the beer and bread was dictated by the grains that each society could grow in the environment in which they lived. Beer offered the mood-altering properties that seem to be sought by humans as well as a much safer beverage than water. Men, women, and children consumed beer with each meal and between meals. Bread offered another form for grain consumption requiring the same basic grain ingredients. As pointed out by Tom Standage (*A History of the World in 6 Glasses*), bread and beer "were different sides of the same coin: Bread was solid beer, and beer liquid bread".

## Sources

Hornsey IS (2003) *A History of Beer and Brewing.* The Royal Society of Chemistry, Cambridge, UK.

Janick J (2002) *Agricultural Scientific Revolution: Mechanical, Figure 21.* [Online] Available at: http://www.hort.purdue.edu/newcrop/history/lecture32/h_21.html [Accessed 26 January 2010].

Kiple KF, KC Ornelas (2000) *The Cambridge World History in Food, Volume One.* Cambridge University Press, Cambridge, UK.

Mares W (1984) *Making Beer.* Alfred A. Knopf, NY.

Standage T (2005) *A History of the World in 6 Glasses.* Walker & Co., NY.

# 5

# Human Nutrition and Diet

*What makes your heart feel wonderful,*
*Makes (also) our heart feel wonderful.*
*Our liver is happy, our heart is joyful (...)*
*I will make cupbearers, boys (and) brewers stand by,*
*While I turn around the abundance of beer,*
*Drinking beer, in a blissful mood,*
*Drinking liquor, feeling exhilarated,*
*With joy in the heart (and) a happy liver.*

Sumerian Drinking Song
Third Millennium BCE

The transition from the hunter-gatherer life style to diets based virtually only on grain consumption had profound consequences for human health. Hunter-gatherers took advantage of the changing availability of various animals and plants at differing locations and times of the year. The evolution of the human body was influenced by hundreds of thousands of years of the richness of an omnivore diet. This diet was based on a range of foods providing nutrients from a number of sources resulting in

© T.R. Sinclair and C.J. Sinclair 2010. *Bread, Beer and the Seeds of Change:* *Agriculture's Imprint on World History* (Thomas R. Sinclair and Carol J. Sinclair)

relatively good growth and health. Critical amino acids, vitamins, and minerals were provided abundantly by consuming a wide mixture of foods. In fact, humans are nearly completely dependent on external sources for several specific amino acids and vitamins since the human body cannot synthesize these nutrients. Failure to consume these essential nutrients that cannot be readily synthesized in the body results in serious health problems.

In contrast to hunter-gatherers' bounty, a very small fraction of people living in agrarian societies sustained rich omnivore diets. Perhaps the Bantu in Africa were able to consume meat on a regular basis, but in most societies meat was available consistently to only the elite. Everyone else existed on a grain-based diet – beer and bread. Overall, a grain-based diet can result in chronic health problems including increased infant mortality, incidence of infectious disease, dental problems, and shortened life span. One distinctive contrast between agrarian societies and hunter-gatherer societies was the difference in bone structure and height. People were smaller in ancient agrarian societies than either the hunter-gatherers that preceded them, or those who enjoy the varied diet of modern times.

Another major change in the diet in transitioning to agriculture was the substantial increase in caloric intake needed by those who grew crops. Human muscles provided the energy required in the intensive labor of growing crops. The fuel for these muscles was necessarily the caloric intake of the workers. High food intake in agricultural societies was essential to undertake the work needed in preparing fields for sowing, digging irrigation canals, weeding fields, harvesting, and threshing. A legacy left to modern humans is the appetite to consume large quantities of calories. Today, without the hard labor to burn these calories, weight gain and obesity result.

The need for food intake is directly related to the amount of physical effort. Hard, physical labor of growing crops resulted in a large requirement for input calories. Modern Recommended Daily Allowance (RDA) gives the basal caloric requirements for humans of differing gender, age, and weight. For example, men weighing 85 kilograms (187 pounds) in the 18-to-30 year age bracket and in the 30-to-60 year age bracket are estimated to have a resting energy expenditure of 1980 and

1865 kilocalories per day, respectively. A woman weighing 65 kilograms (143 pounds) in each of these age brackets has a resting energy expenditure of 1451 and 1394 kilocalories per day, respectively.

To estimate actual dietary energy needs, these basic energy expenditures for resting are increased based on the level of physical activity. A multiplier is used to account for the level of activity and the number of hours the activity is sustained each day. For example, very light physical activity such as sitting, driving, typing, or cooking is accounted for by a multiplier of 1.5 for the hours in which a person engages in this activity. Assuming a person is at rest for half the day and involved in very light physical activity for the other half, which is probably common in modern times, the daily multiplier would be about 1.25 [(1.0 + 1.5)/2]. Therefore, the average general estimate for required caloric intake for men is roughly 2400 kilocalories per day and for women is roughly 1800 kilocalories per day. The challenge in modern diets is often to keep the caloric intake down to these levels.

On the other hand, people in societies prior to modern times were involved in the heavy labor of growing and processing crops. The activity multiplier for even moderate levels of activity is at least 5.0 for those hours worked. Since the labor in historical crop production continued from sun up to sun down, we can assume that the multiplier of resting energy expenditure may have been about 3.0 [(1+5)/2] for those using their muscles to produce food. The 3.0 multiplier is, of course, more than double the value of 1.25 estimated for modern people.

Since the stature and weight of people in historical societies were less than modern people, a downward adjustment can be made to the base energy needs of these people. Therefore, their resting energy expenditure would have been less. The caloric requirement for an 18-to-30 year old man weighing 65 kilograms (143 pounds), using the 3.0 multiplier to reflect the heavy work of growing crops, is estimated at 5020 kilocalories per day. Even with smaller bodies, men in past societies are estimated to need more than twice the caloric intake of our modern needs. Calculation for the 30-to-60 year old group is of less significance because of the limited number of adults that previously survived into this older age bracket.

Women were fully engaged in the production of food, and their energy needs would have also been high. An 18-to-30 year old woman weighing 50 kilograms (110 pounds) doing moderate labor would have a caloric requirement of 3693 kilocalories per day. Further, in past societies women were commonly pregnant or lactating much of the time during their reproductive years. Pregnancy adds to the daily energy requirement for a woman. The estimated additional caloric intake for a pregnant woman is on the order of 300 kilocalories per day and for a lactating woman on the order of 600 kilocalories per day. Therefore, for much of their adult lives women in these earlier societies needed over 4000 kilocalories per day.

The diets of people in the historical agricultural societies considered in this book were commonly dominated by one or two cereal grains. The calories were obtained from carbohydrates – starch and sugars – stored in these grains. The total amount of energy stored in carbohydrates is about 4 kilocalories per gram. Therefore, a ballpark calculation indicates that people needed to consume at least 1000 grams (2.2 pounds) of grain daily to obtain 4000 kilocalories. When the grain was consumed as less energy-dense bread or beer, the actual weight consumption required was greater. As bread, the caloric intake is roughly about 2 kilocalories per gram. Therefore, over 50 slices of modern bread (80 kilocalories per slice) would be needed to obtain the minimum required 4000 kilocalories per day.

Bread, of course, was not the only source of calories; fermented grain accounted for a large part of the caloric intake. The actual caloric content of the beer would have varied a great deal depending on the fermentation process, whether dissolved materials were filtered out or consumed, and alcohol content. If we assume approximately 50 kilocalories per 100 milliliters (slightly less than half a cup), which is common for modern beers, to obtain 4000 kilocalories a person would need to drink 8 liters (2 gallons) of beer each day. Since most ancient beers were not filtered, it is likely that the caloric content was much greater than the estimate we used based on modern filtered beers. Nevertheless, it is not surprising that beer was consumed in large quantities at all meals by everyone to help sustain the needed caloric intake. It was, in fact, the only beverage consumed by most people for most of their lives after weaning from mother's milk.

The large daily consumption of beer, including even pregnant women, raises questions about the prevalence of fetal alcohol syndrome among historic populations. Fetal alcohol syndrome results from exposure of a fetus to alcohol and is associated with a wide range of structural and biochemical alterations in the developing brain. The effects of the syndrome range from minor changes in neurobehavior to mental retardation and craniofacial deformities. In modern times, there is a clear recognition that alcohol consumption by women during pregnancy can result in serious consequences for the fetus. For example, in the U.S. bottles of beer are required to have labels warning of the risk of birth defects. However, recognition of this problem is not new. Aristotle said "foolish, drunken ... women must bring forth children like unto themselves." In some societies children of women known to be regularly drunk were cast away and abandoned.

With the modern recommendation of complete abstinence from alcohol during pregnancy, it would seem that fetal alcohol syndrome must have been serious and widespread in historic societies given the fact women may have been consuming 1.5 to 2 liters (3 to 4 pints) of beer per day. Using modern statistics it is possible to roughly estimate the severity of fetal alcohol syndrome. The most severe effects of alcohol are associated with average daily alcohol consumption of 90 milliliters (3 ounces) or more. For beer of 3% alcohol, which is likely the range of much beer in history, a woman would have to consume 3 liters (6 pints) per day, well above what was likely routine consumption, to cause severe fetal alcohol syndrome. In addition, modern surveys to establish the threshold for severe fetal alcohol syndrome were based on weekly alcohol consumption, and it is likely that consumption by these modern women is concentrated on weekends, which subjects a fetus to large spikes in exposure to alcohol. Therefore, it seems likely severe fetal alcohol syndrome was not common through history.

However, consumption of 1.5 to 2 liters of beer by women through much of history is markedly different than the modern recommendation of total abstinence. In fact the modern recommendation includes a safety margin beyond the statistical threshold for measurable fetal impact of one drink per day (15 milliliters alcohol, or 0.5 ounces), which has been shown to result in adverse neurobehavior effects. This limit is equivalent to consumption of 0.5 liters (1 pint) per day. Based on these modern findings, fetal alcohol syndrome at some level of limited cognitive processing seems likely through much of history. Of course, the consequences of mild expression of the fetal alcohol syndrome would have had little impact on lives filled with the drudgery of growing crops and preparing bread and beer.

There were additional health consequences for those throughout history whose diets consisted of large intakes of bread and beer derived from cereal grains. In addition to consuming kilocalories to provide energy for the body, it is necessary to consume essential amino acids for construction of proteins. Protein is an essential input of the human diet simply to sustain the biochemistry of life. Virtually all the functions of the human body operate through the activity of proteins. Therefore, it is essential to have a diet that provides sufficient protein to replace the daily turnover of proteins in the human body. The basic amount of protein needed by humans per unit of body weight is fairly stable across gender. The RDA estimate of daily protein needs based on daily loss in excretion is approximately 0.75 grams per kilogram body weight. Therefore, the man and woman from agricultural societies for whom caloric intakes were estimated above would have protein requirements of approximately 50 and 40 grams per day, respectively. Most cereal grains contain 8 to 20% protein. In addition to the protein obtained directly from the grain, yeast in the beer consumed can result in a boost in the amount of dietary protein. Therefore, a diet based on at least 1000 grams of daily grain consumption – a large portion of which is yeast-enriched beer – would readily provide the required amount of total protein.

A complication in the protein intake is that the proteins also need to be composed of the proper combination of amino acids to meet the needs of the human body. Unfortunately the amino acid composition of the proteins in cereal grains does not match well with human requirements. Humans need to ingest eight essential amino acids that are not synthesized within their bodies: phenylalanine, valine, threonine, tryptophan, isoleucine, methionine, leucine, and lysine. Meat from animals, which have similar metabolism to humans, usually provides these amino acids, while a diet based almost solely on cereal grains does not. Plants have metabolic pathways and biochemical requirements different from those of mammals for storage of proteins in their grains; therefore it is not surprising that the amino acid composition of grains does not necessarily match well with human requirements.

Amino acid deficiencies vary substantially among the grains of cereal species. Oats, for example, actually has an amino acid balance that matches reasonably with human requirements. However, most of the other cereal grains are deficient in lysine, tryptophan, methionine, cystine, and histidine relative to human needs. Maize, in addition, is deficient in isoleucine. Lysine, in particular, is a crucial amino acid that is lacking in a diet limited to direct consumption of cereal grains. Only oats and rye have reasonable levels of lysine; millet, sorghum, and maize are especially low. However, some yeasts can synthesize lysine during fermentation of grains to produce beer and can result in a natural supplementation of this amino acid. As with other amino acids, lysine is essential in building the proteins needed in the human body. Without an adequate supply of lysine for protein synthesis, growth and development are retarded in children. The deficiency symptoms include poor appetite, low activity level, and slow healing of wounds. Adults also can suffer some of these same problems.

Cereal grains also do not necessarily provide the essential vitamins and minerals needed by the human body. Serious diseases can result from insufficient intake as a result of a concentrated reliance on cereal grains. Vitamins A, $B_{12}$, and C are not available in many cereal grains. In modern societies, fortified dairy products are sources of vitamin A and $B_{12}$; fresh fruits and vegetables are sources of vitamin C and beta-carotene, which is a precursor of vitamin $B_{12}$. Such varied diets did not exist for most people in historical societies. Even at best, only small quantities of fruits and vegetables rich in vitamin C would be available from gardens. Therefore, deficiency symptoms for these three vitamins would likely have been prevalent through much of history. Vitamin A deficiency, which is especially prevalent in young children, results in ocular problems and susceptibility to infections. Vitamin $B_{12}$ deficiency causes irreversible neurological damage, which results in cognitive dysfunction. Vitamin C deficiency results in scurvy, which is associated with widespread capillary hemorrhaging. These diseases would surely have been found in most societies, except perhaps in more tropical environments where the diet could have been more readily augmented with collected vegetables and fruits.

Cereal grains do provide several vitamins including $B_6$, folate, pantothenic acid, and E, although at low concentrations. Consumption of large quantities of grain would have provided sufficient intake to meet the daily recommended dietary allowance. However, intake by itself does not reflect the nutritional complications for deficiencies of these vitamins. Cereal grains contain compounds that prevent vitamin absorption during digestion, or depress the nutritional benefit of individual vitamins. Niacin is an important vitamin whose absorption is minimized due to the presence of anti-nutrients in the grain. Niacin is available in quantity from meat and can be synthesized in the human body from the amino acid tryptophan. While cereal grains have reasonably high levels of niacin, the anti-nutrients in these grains cause niacin to be unavailable for intestinal absorption. Fermentation of grains can result in a substantial improvement in the niacin supply since yeasts synthesize niacin for their own growth. Without adequate niacin absorption, pellagra develops. Estrogen adds to the problem of pellagra: there is a 50% higher incidence of pellagra among women. Pellagra is characterized by the 4 Ds: dermatitis, diarrhea, dementia, and death. Pellagra epidemics were recorded as recently as the $20^{th}$ century in the United States among the poor as a result of high consumption of canned hominy grits. As discussed in Chapter 11, preparation of maize grain for fermentation, as done by the ancient Maya, increased the bioavailability of niacin.

Cereal grains are also low in quantities of some minerals, including calcium and sodium. Calcium is, of course, needed for developing and sustaining bones. Populations dependent on cereal grains surely suffered osteomalacia, which in children is identified as rickets, and also osteoporosis. Sodium is an important electrolyte in the human body. While in modern salt-rich diets sodium is generally consumed in excess, in some ancient societies salt was in short supply.

All minerals except calcium and sodium would have been obtained in abundance as a result of the large amounts of grain consumed. Like some of the vitamins, however, absorption of some minerals from grains is very poor in human digestion. It appears high levels of phytate in grains significantly limit bioavailability of these minerals. Phytate combines with phosphorus in plants as a mechanism to store phosphorus. However, animals without rumens, including humans, cannot digest the phytate-phosphorus complex. Phytate can also combine with iron, zinc, calcium, and magnesium which decreases their absorption. Fortunately for beer consumers, yeast growth during fermentation decreases the concentration of phytate.

---

**Box 5.1.** Salt

An essential mineral for human health, salt was in demand as an important commodity of commerce in ancient societies for several reasons. Salt added to gruel would improve taste. In addition, at times when large animals were slaughtered, meat that was not immediately consumed had to be preserved. In some societies, there would not be enough feed to carry all animals through the winter so many of the animals were therefore slaughtered in the fall. Animal products were salted to keep them from rotting so they could be rationed through the winter.

Salt was important in the history of several ancient societies. Jericho, sometimes identified as the world's oldest city, probably developed as a salt trading center considering its proximity to the Dead Sea. Water from the Dead Sea left to evaporate in ponds yielded large quantities of salt. Ostia, the port through which grain flowed to Rome, also had evaporation ponds to provide salt. The European cities of Venice and Salzburg (literally salt castle) trace their roots as salt trading centers. Taxes imposed on salt through history have been a source of conflict between governments and their subjects.

---

Bioavailability of iron and zinc is a particularly serious problem for those on diets of virtually all grain. Iron deficiency leads to anemia, which is a disease afflicting many in the modern world whose diet contains little meat. Anemia in adults results in decreased activity and capacity to work. Anemia is especially severe in infants who lack iron and leads to increased pre- and post-natal mortality. Those children who survive the anemia often suffer irreversible impairment of learning ability. Zinc deficiency results also from its limited availability from grains. The symptoms of zinc deficiency include growth retardation and dwarfism.

Overall, a diet focused solely on the consumption of large amounts of grain would provide the energy and protein required for human muscles to perform the heavy workloads common in historic societies. However, reliance on grains would have resulted in serious nutritional problems from lack of bioavailability of vitamins and minerals. These dietary deficiencies would have resulted in high infant mortality, inhibited growth and brain development of children, and vulnerability of adults to diseases. Specific health problems would have included scurvy, pellagra, rickets, and anemia. It is little wonder that pre-modern agricultural societies have been found to be less healthy than the early hunter-gatherer populations who preceded them.

---

**Box 5.2.** Beriberi

Beriberi is one nutritional disease associated with a grain diet that has not been included in the discussions of this chapter. Beriberi is a result of a deficiency of vitamin $B_1$, which is also identified as thiamine. The disease is associated with a number of symptoms including loss of appetite and weight, nausea, nervousness, and fatigue. Interestingly, whole cereal grains contain sufficient thiamine to prevent beriberi, but the thiamine is contained in the outer layers of the seed. When the Chinese, in particular, chose to eat polished rice, the removal of the outer seed layers in polishing resulted in a loss of thiamine and a rise in the incidence of the disease. Since rice of any type became a major component of the Chinese diet only after the Golden Age discussed in this book, beriberi was not a problem for this society during its Golden Age discussed in Chapter 9.

## Sources

Brenton, BP (2004) *Piki, poletna, and pellagra: Maize, nutrition, and nurturing the natural.* In: R. Hosking (ed), Nurture. Proceedings of the Oxford Symposium on Food and Cookery 2003, Footwork, Bristol, UK.

Cordain L (1999) Cereal grains: Humanity's double-edged sword. *World Review of Nutrition and Dietetics* 84:19-73.

Henry RH, PS Kettlewell (1996) *Cereal Grain Quality.* Chapman & Hall, London, UK.

Jacobson, JL, SW Jacobson (1994) Prenatal exposure and neurobehavioral development: where is the threshold? *Alcohol Health & Research World* 18:30.

National Research Council (1989) *Recommended Dietary Allowances 10th Edition.* National Academy Press, Washington, DC.

Symons M (2000) *A History of Cooks and Cooking.* University of Illinois Press, Urbana and Chicago.

# 6

# Cropping 101: Water, Nitrogen, Weed Control

*When you plant lettuce, if it does not grow well, you don't blame the lettuce. You look into the reasons it is not doing well. It may need fertilizer, or more water, or less sun. You never blame the lettuce.*

Thich Nhat Hanh

*Even the richest soil if left uncultivated will produce the rankest weeds.*

Leonardo da Vinci

Anyone who grows a vegetable garden learns quickly that seeds cannot simply be sown in the soil and ignored until harvest. The garden needs preparation and attention throughout the season to ensure that the plants have a good environment for growth. Two key resources are water and nutrients, especially nitrogen. Water must be provided regularly to young seedlings in the garden to ensure their establishment. Later in the summer when days become hot and dry, irrigation may be needed in quantity to keep the plants from wilting and dying. Similarly, nutrients must be available in garden soil so plants grow readily and are productive. Crop plants need water and nutrients to ensure good yields. Indeed crop yields are closely linked to the amount of water and nitrogen available to crops. The ability of farmers in antiquity to provide the water and nitrogen needed by their crops had a large influence on the course of history. In this chapter we examine the reasons that water and nitrogen must be available in abundance to crops.

The evolutionary transition of plants from lakes and seas to land masses succeeded because both sunlight and carbon dioxide were more abundantly available. Leaves evolved giving plants large surfaces to intercept sunlight and to absorb carbon dioxide in the air. Photosynthetic rates of land plants were much greater than had been achieved by plants submerged in water. However, land plants were faced with an atmosphere that was bone-dry relative to the water content of the plant cells. No large plants could develop without mechanisms to protect themselves from rapid dehydration due to water evaporation into the atmosphere.

One of the evolved traits to protect land plants from desiccation was to cover leaf surfaces with a waxy material. These waxes form a cuticle, a barrier nearly impermeable to water vapor. A problem with a continuous cuticle, however, is it also prevents diffusion of carbon dioxide from the atmosphere into the interior of the leaf where photosynthesis takes place. The solution to this dilemma of obtaining carbon dioxide but minimizing water loss was the evolution of micro-pores in the leaf surface controlled by the plant. These pores, called stomata, open when conditions are suitable for photosynthesis such as during daylight hours (Fig. 6.1). The "penalty" of opening the stomata, of course, is that the gas pathway for carbon dioxide is the same pathway used by water molecules to diffuse from inside the leaf out to the atmosphere in a process called transpiration. Not surprisingly, there is a very close relationship between the rate of photosynthesis, i.e. growth, and the rate of transpiration. To have stomata open for high photosynthetic rates and high plant crop growth, there is no alternative but for large amounts of transpired water to move from the leaves to the atmosphere.

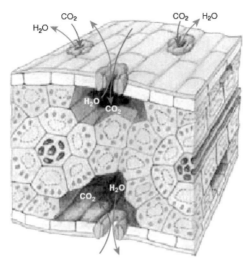

**Figure 6.1.** Drawing of a cross-section of grass leaf. Carbon dioxide for photosynthetic assimilation diffuses from the atmosphere into the leaf through the stomatal pores in the leaf surfaces. The open stomata also allow water vapor to diffuse from inside the leaf to the atmosphere.

The physics of the gain in carbon dioxide by plants and the transpiration of water into the atmosphere is well resolved. The relationship depends on the humidity of the atmosphere, but a ballpark figure for a "representative environment" is that the accumulation of about 1 gram (0.035 ounces) of plant mass through photosynthesis will result in roughly 500 grams (1 pound), or about 0.5 liter (1 pint), water loss via transpiration. Hence, a crop that has accumulated through the season 400 grams of plant matter per meter square – a representative crop growth through much of history – would be associated with water loss in the range of 200 liters of water per meter square. Converting this water volume per unit land area to a depth of water, the crop would lose directly a water depth of 200 millimeters (approximately 8 inches) as it grows. However, this transpiration from crop plants usually accounts for less than half of the total water loss from a field. Considerable water can also be evaporated directly from the soil surface under the plants. Weeds also transpire water and can account for substantial water loss. Providing crops with sufficient water so the stomata can remain open and allow photosynthesis through the growing season is a major challenge in sustaining production in a number of environments.

Supplying water for crops was a fundamental challenge in maintaining food supplies in many societies. Crop growth on flood plains, such as practiced in Sumeria and Egypt, was possible because rivers provided the water needed by their crops. Societies that were solely dependent on rainfall were especially vulnerable to drought. The ancient Greeks and Romans suffered soil erosion to the point there was little soil left in which to store rainfall. These two societies eventually lost their ability to produce grains in sufficient quantities to feed themselves and needed to bring in grain from conquered territories. Mayan society appears to have dissolved when a succession of dry years resulted in crop failure.

Water alone is not sufficient to ensure a good crop yield. The amount of nitrogen required by plants (and nearly all other living organisms) imposes by far the most rigid limitation among nutrients that must be accumulated within crop plants. Nitrogen is required as an essential constituent in the proteins that are the biochemical engines doing the plant's work. In addition, nitrogen is an essential component of the nucleic acids used to construct DNA and RNA – the genetic memory of all living cells. It is impossible for a cell in plants, or in humans for that matter, to develop and survive unless it has the required amounts of protein and nucleic acid. Low amounts of nitrogen result in stunted growth and yellowing of the plant. There is no alternative to good crop growth other than accumulation of large amounts of nitrogen by plants.

The amount of accumulated nitrogen needed to achieve growth is well defined. For example, a wheat crop contains about 1.5% nitrogen. Therefore, a wheat crop accumulating 400 grams of plant matter per meter square, the example used previously, must accumulate approximately 6 grams of nitrogen per meter square. While this is a small amount compared to the quantity of water needed, nitrogen is scarce and ephemeral in the natural environment. Even though plants are surrounded by nitrogen in the atmosphere, atmospheric nitrogen cannot be used by most plants. Instead, nitrogen – usually in inorganic form – must be recovered from the soil by plant roots.

The natural environment offers only a few sources of nitrogen. One source is lightning storms in which the heat and energy in lightning bolts convert nitrogen gas in the atmosphere to chemical forms that are washed into the soil with rain water. Once in the soil these nitrogen forms can be readily absorbed by plants. In regions where thunderstorms are fairly frequent, there may be as much as 0.1 to 0.3 grams per meter square of available nitrogen deposited on the land surface per year. Another source of nitrogen for plants is microbes living in the soil that can metabolize atmospheric nitrogen directly to synthesize their proteins and nucleic acids. After the microbes die in the soil, the nitrogen they had in their bodies becomes available for plant growth. There are, however, a number of constraints on these microbes including their needs for appropriate soil temperature and soil water content. The nitrogen input to the soil from these microbes under good conditions is usually no more than the input from lightning. Of course, in arid regions with little rain and few thunderstorms and with dry soil, both of these sources of nitrogen input are near zero.

Natural ecosystems persist when they can sequester the nitrogen in live or dead plant tissue. In this way, over many years ecosystems can accumulate nitrogen as standing plant material and as plant residue in the form of organic matter in the soil. Once the natural ecosystem is disturbed and the nitrogen is lost, it can take many years to recover nitrogen levels in the ecosystem to levels that existed prior to the disturbance.

The challenge in cropping is that much of the available nitrogen is lost from the field each year when harvested plant material is removed. There is no ability to sustain crop production unless sources of nitrogen are regularly added to the soil. Until very recent times, the main source of nitrogen for crops was from organic matter either in the soil as plant residue, or added as animal manure. Soil microbes metabolize these various sources of organic matter and eventually release the nitrogen into the soil in chemical forms that can be readily taken up by plant roots. Until very recent times when manufactured nitrogen fertilizer became available, providing organic sources of nitrogen for cropping was a critical factor in sustaining cropping. Societies that grew crops on flood plains depended on the organic matter imported from ecosystems upstream, which was a key advantage in Sumerian and Egyptian cropping. Later, societies learned to regenerate organic matter within the cropping system by using crop rotations that included legumes. Legumes have the unique ability to form symbioses with specific bacteria in a way that allows gaseous nitrogen to be metabolized into forms that can be used by plants. Legumes play a critical role in some natural ecosystems because of this ability to "fix" atmospheric nitrogen. Agricultural rotations that included a legume, when introduced in Great Britain beginning about 1700, led to a doubling in cereal grain yields. This tremendous increase in crop productivity profoundly affected the social order of Great Britain. The number of people needed on the manors to grow crops correspondingly decreased and a work force became available for the new factories of the emerging Industrial Revolution. In contrast, the Bantu societies of Africa solved their nitrogen limitation by moving to new lands when the organic matter in their current fields was exhausted. The virtually limitless land area in eastern and southern Africa made the migration from old fields to new fields possible.

A major consequence of providing crop plants with water and nutrients for good growth is that the soil environment is also made very attractive for other plants. Some native plants are especially responsive to an improved soil environment developed for crops. These unwanted plants in the crop fields are weeds that compete with crop plants for water and nutrients. Water and nutrients consumed by weeds are lost to the crop plants and yields suffer. If weed plants become sufficiently tall, they shade crop plants and dramatically decrease crop yields. Also, large weed plants complicate harvest and could adulterate the harvested grain. The bane of growers throughout history has been the nearly continuous and tedious work of removing weeds from their crops. In fact, many of the onerous tasks in growing crops can be traced to efforts to minimize weed infestations. The struggle to diminish the influence of weeds can begin well before the crop is sown and usually continues throughout the life cycle of the crop.

Weeds that earn the special wrath of farmers are those that have evolved mechanisms for long-term persistence in crop fields. One mechanism for a weed species to persist is to produce huge quantities of seeds. A large "seed bank" develops in the soil staying dormant until conditions are ripe for germination and seedling establishment. A flush of new weeds will emerge, for example, when a heavy rain wets the soil. Commonly, weeds have also evolved mechanisms to ensure non-uniform seed germination so that a new crop of weed seedlings can continually emerge throughout the growing season. Non-uniform germination can occur when weed seeds have delaying mechanisms that result in a range of germination dates for the populations of weed seeds in the seed bank. Sometimes a population of weed seeds will include some seeds that are viable for several years.

Weed persistence also results when below-ground organs are produced that can generate a number of daughter plants. Hoeing may remove the top of one of the daughter plants, but the organ for producing another daughter plant remains unharmed. Nutsedge is one of the most virulent weeds worldwide because it produces tubers on the root that can generate four to six successive daughter plants. Simply removing the first few flushes of nutsedge plants above ground offers no lasting control of this weed.

Much of the technological development throughout the history of agriculture has been focused on controlling weeds. Hand hoeing was the earliest control and, of course, is still used in many places in the world in growing crops and in the vegetable garden. Historically, implements and cropping systems were developed that helped in the effort to suppress weeds. In very recent times, specific chemicals, i.e. herbicides, were discovered as discussed in Chapters 15 and 16 that attack specific weed species with little or no impact on the crop plants.

**Sources**

Sinclair TR, FP Gardner (1998) *Principles of Ecology in Plant Production.* CAB International, Wallingford, UK.

# 7

# Sumerians
# ~3500 to 2334 BCE

*A harlot introduces to Enkidu the pleasures of civilization, including the eating of bread and drinking beer. After eating his fill and becoming drunk on seven jugs of beer, Enkidu took a bath, anointed himself with oil and "became human".*

Ancient Sumerian Story of Gilgamesh

The Fertile Crescent was one of the earliest, if not the first, areas where humans domesticated plants and grew crops. This area stretched from the shores of the eastern Mediterranean Sea, north to the slopes of the Taurus Mountains, and arched southeast along the Zagros Mountains. Emmer wheat (*Triticum turgidum*) and barley (*Hordeum vulgare*) were native grasses to the region and were harvested by the early hunter-gatherers. Importantly for the development of the Sumerian society, the Euphrates and Tigris Rivers flowed through the arid region west of the Zagros Mountains forming the large, flat flood plain of the Mesopotamia Valley (Fig. 7.1). The flood plain of these rivers was free of stones and rocks, and the soil could be easily tilled. Seeds of the native wheat and barley were sown on the Mesopotamia flood plain by the first agriculturalists.

Capturing water from the rivers was key to producing abundant crops in this arid region.

By about 3500 BCE, techniques for producing wheat and barley had been fully developed to feed a growing population living together in several cities. These city-states, collectively identified as Sumeria, developed along the southern extent of the Euphrates River (Fig. 7.1). The city-states were the first full-scale societies with class structure to carry on the various activities of "civilization", including growing crops. A small ruling class arose that organized the labors of the great majority of the rest of the people to assure the production of food. From the Golden Age of Sumeria emerged the first organized governments, writing, and mathematics – all necessary for the administrative control in crop production.

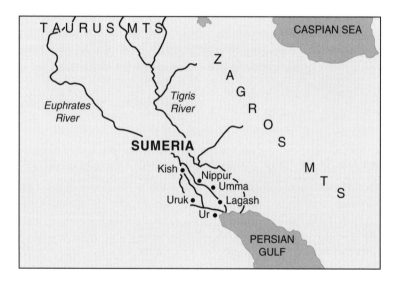

**Figure. 7.1.** Map of Mesopotamia region with Sumeria located in the area of the southern part of the Euphrates River. Major Sumerian city-states are identified. (Redrawn from Smitha (n.d.).)

The rivers' water and flood plains, which were the base of sustained food supply, developed over millennia. The Tigris River originated in the Zagros Mountains of western Iran and in the Taurus Mountains in the north. The Euphrates originated farther to the west in the Taurus Mountains and had a longer path through arid lands before reaching the southern regions of the Mesopotamia Valley. As these rivers flowed to the south out of the mountains they carried with them the silt and organic matter scoured from these slopes. The key to agriculture in this region was the annual flooding by the rivers leaving a deposit of silty soil and organic matter. The relatively flat, broad flood plain made an especially attractive area to grow crops. By 3500 BCE, fields of wheat and barley were being grown to feed the population of the Sumerian civilization. In contrast to modern Iraq with a society based on oil production, the Sumerian society was based on water and silt delivered by its rivers.

Major challenges, however, existed in developing a viable cropping system in this environment. Wheat and barley must be grown when temperatures are cool; these plants are unable to grow and yield satisfactorily in the hot arid summer. Low yields under high temperature result because the life cycle of these plants is greatly accelerated and only a few grains are produced, and these grains may not even be viable as seeds for the next crop. To avoid the high summer temperatures, it is necessary to grow these plants during winter months. However, the timing of when the crops could be grown was out of synchrony with the stages of water flow in the rivers. The rivers were largely fed by winter rains that fell in the mountains. The peak flow of water in the Euphrates would usually reach the lower Mesopotamia Valley sometime in April, just when scorching temperatures were building in the region. No crop would be waiting to absorb the readily available waters during the peak of flow, and efforts had to be focused on growing crops after the lands had been dried by the summer heat. Crop production was only possible by developing an irrigation system to bring water from the river to the fields during periods in the year when river flow was low.

The critical achievement of the Sumerians was the development of a canal system that controlled the flow of water to the fields. Since energy requirements to lift water from the river to the fields could be great, the

Sumerians constructed a system dependent on gravity. The rivers had to be tapped upstream, at a point where the elevation of the river was greater than the fields, so the water could flow down through canals to the fields. Main canals were built to tap into the rivers and carry the water to grids of secondary and tertiary canals that carried water to individual fields. Maintenance of this canal system was essential in sustaining food production.

The Tigris and Euphrates Rivers were not equally useful for irrigation. The Tigris was partially fed from the nearby Zagros Mountains, which resulted in vigorous flow during the winter rains cutting a deep river channel. Consequently, the deep channel of the Tigris made it a difficult source for irrigation. To achieve the necessary gravity gradient from the river to the fields, an irrigation canal would have to be cut far upstream. Since canal construction was done by human muscle, it surely would have been attractive to keep the length of the main canal as short as possible. The Euphrates River originated in the mountains far to the north and flowed over 2700 kilometers (1700 miles) from the Taurus Mountains. Much of the course of the Euphrates was through arid regions where evaporation from the river would have removed a substantial amount of its original water, especially during the summer months. Daniel Hillel (*Out of the Earth*) estimated that about half of the water leaving the mountains actually reached lower portions of the river. By the time the river reached the lower Mesopotamia Valley it was a sluggish river with high silt content. Instead of scouring a deep river channel, silt would be deposited on its banks and in its channel. The accumulation of silt resulted in a slow increase in the elevation of the river. Consequently, short irrigation canals from the Euphrates River became the major source of water for irrigating the grain fields, and most Sumerian city-states developed in the vicinity of the Euphrates River.

A major impetus to sustain these city-states with a ruling class and a large working class must have been the need to construct and maintain the large canal system. The high load of silt in the river would settle in the irrigation canals. The effort to remove silt from all canals and keep water flowing to the fields was never ending. A powerful ruling class was necessary to coerce the vast majority of people to labor in the many tasks of producing food. The invention of writing in roughly 3300 BC would have been one of the skills needed by the ruling class to organize this first large-scale agricultural society.

The water of the Euphrates irrigated the crops and also maintained the fertility of the fields. Organic matter eroded from the Taurus Mountains would be suspended in the irrigation water and carried onto the fields. The continual resupply of organic matter brought in with irrigation of the fields gave a supply of materials to the microbes that released nitrogen needed by the crops.

No doubt, the Euphrates also brought to the Mesopotamia Valley steady supplies of weed seeds that were carried to the fields in the irrigation water. Each irrigation of the fields would have also resulted in the sowing of a large number of weeds. In the spring after crop harvest, weed seeds would have sprung to life if water was available in the crop fields. Water seepage from the river and canals through the soil could have supplied sufficient moisture deep in the soil, at least for the early part of the summer for weeds to grow. Weed growth during the summer would have consumed the nitrogen released by the microbes in the field. Also, the fields would have been left choked with plant residue that would have made sowing the subsequent crop very difficult. Therefore, in addition to the challenging task of cleaning the irrigation canals of silt during the summer months, weeds had to be cleared from the fields.

The Sumerians developed a simple wooden implement used in attempts to control weeds. This simple tool, called an ard, is commonly identified as a plow. The implement probably was pulled by one individual with a second holding the handles to guide the ard and maintain its depth in the soil (Figure 7.2). The task of pulling the ard would have been physically demanding, but possible because the soil was friable and free of stones and heavy clay. Stone carvings showing the ard being pulled by oxen have been found. However, it is likely that this was not a common approach because feeding of such large animals year round would have been very challenging. During summer months, in particular, when fields needed to be tilled, there would have been little fodder available to feed large animals.

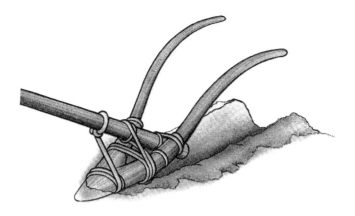

**Figure 7.2.** Drawing of ard for tilling soil. Likely, men often pulled this hoe through the friable soil of the Euphrates flood plain. (From Burke 1978, courtesy of James Burke.)

It is probably more accurate to describe the ard as a hoe rather than a plow since it would disturb only the top soil layer. Scratching the soil surface would, of course, be sufficient to kill young weed seedlings emerging from the soil. The soil would have been repeatedly "plowed" to kill the young weed seedlings as they emerged from the soil. Young seedlings still have shallow roots and are not yet established in the soil. Once large weed plants became established, they were very difficult to remove and left a residue of plant material to interfere with sowing the subsequent crop even if the weeds eventually died during the arid summer heat. Each pass through the field with the ard would also have caused the top soil layer to dry quickly resulting in a hot, dry zone of soil potentially killing weed seeds and also the germinating weed seedlings.

At the end of the hot summer, it was time to prepare the fields for sowing the crops. Fields had to be flooded before sowing to replenish the soil with water that had evaporated in the long, arid summer. In October/November, the irrigation of the fields would begin. Water flowed through the canals to flood each field. To maintain water on the fields until it had a chance to penetrate into the deep soil layers, the fields were surrounded by small dikes. During the long, hot summer these dikes had to be constructed and repaired. Re-wetting of the fields must have been a time of intense effort to manage the distribution of water to each of the fields and to quickly repair any breach of the dikes.

Once the water from the flooding of a field had penetrated into the soil, it was time to sow the fields. There is evidence that the Sumerians used a modified ard to sow the crop. There are drawings from the Sumerian era showing an ard to which a funnel was attached that allowed seeds to be dropped in place in the ground behind the "shoe" of the ard. A third person was likely added to the ard team during sowing to drop seeds in the funnel. The use of the ard with a funnel would have resulted in rows of plants, which would also allow use of the regular ard later in the season to hoe weeds between rows. While sowing and subsequent weeding were "arduous" labor, ards would have been effective.

Placement of seeds in the moist soil using the sowing ard would have increased chances for successful seed germination and seedling emergence from the soil. Achieving high establishment rates for the seed was critical because it produced an improved stand of plants for the amount of seed sown. Until the last two centuries in history, crop yield was always expressed as number of seeds harvested per number of seeds sown. Historically, the valuable input to produce a crop was the amount of seed sown – each sown seed was one less seed available to eat. It is only in very recent times that land became the more valuable commodity and yields were expressed per unit land area.

It seems unlikely that the Sumerians practiced monoculture agriculture, in which each field is restricted to a single crop species. Rather, their fields were likely a mixture of barley and emmer wheat, which is an ancestor of our modern bread wheats. Hulls of emmer wheat attach firmly to the seed, so removing the hull during threshing would have been an important task. On the other hand, emmer wheat grain has a high protein content (20% as compared to about 12% for modern bread wheats), which made it a nutritious food source in an environment where there were few animals to be eaten as protein sources. Dilution of bread dough by barley would not have been a serious problem for the high-protein emmer flour. The mixture containing a substantial amount of barley also would have been especially desirable in beer making.

Wheat and barley sown in October/November would have matured in the following March and April. Harvests were done early in Sumerian history by using sickles made from heavily fired clay forming sharp, vitrified cutting edges. However, these sickles shattered easily. They were eventually replaced with wooden sickles that had cutting edges made from imported flint. By the end of the Sumerian era, copper sickles were becoming common.

Grain heads would be cut from the stalks and taken to a central area for threshing. This would likely have been done on an elevated hill where wind could be used to separate the grain from chaff, i.e. winnowing. A floor of flat stone was constructed on which to beat the grain heads to remove the grain. Once grains were released from the heads, the mixture of grain and chaff would be tossed in the air so the wind would blow away the chaff and the grain would fall back to the threshing floor.

After threshing, the grain could be mixed with water and eaten as gruel (Chapter 4). If any grain were cracked during threshing, the gruel would have been more digestible. Still, it would not have been a highly desirable or nutritious product. Among common people onions and other vegetables could be added to the gruel to vary the taste and enhance the nutritional value. On very special occasions meat of sheep or goats may have been added to gruel.

Sumerians consumed considerable amounts of beer, so grain would certainly have been prepared for brewing beer. Barley was prevalent in the harvest grain mixture. The grain was malted (Chapter 4) as the first step in preparing beer. The grain was soaked in water and allowed to remain wet to initiate germination. The grain was kept wet probably no more than a day, and then the malted grain was dried. The drying could have been done effectively on the same stone floor used for threshing. Once dried, the malt was ground and stored.

Grinding of the malt, and the grain for bread, was an especially onerous task. A saddle quern was used for grinding (Figure 4.1). The quern consisted of a saddle-shaped stone on which a rolling-pin shaped rock was rubbed back and forth. The grain was cracked into small pieces and ground into flour. The grinding, invariably the job of women, was done in a kneeling position. Archeological evidence shows that years of grinding resulted in women having bent and arthritic toes and knees along with deformed lower backs. Years of grinding were almost literally back-breaking labor!

The Sumerians were especially fond of beer, the Sumerian word being KAS. It has been estimated that 40% of the barley crop was used in beer production. Beer was an important item of commerce with wages and salaries often being paid in beer. Cuneiform texts give a long list of various types of beers: dark beer, whitish kuran beer, reddish beer, excellent beer, beer (mixed) of two parts, beer from the "Nether world", beer with a head, beer which had been clarified, beer for the sacrifice, and beer for the main (divine) repast. In Sumeria, beer was rarely clarified but rather drunk directly from the brewing vat. Since considerable solid material would be floating in the brew, particularly hulls from the grain, Sumerians drank from the bottom of the vat using "straws" (Fig. 7.3). Several people could sit around a vat sipping beer from individual or shared straws. Originally reeds were used as straws, but among the wealthy straws were made increasingly from valuable metals. No examples of reed straws have survived, but metal drinking tubes were found at the tomb of Queen Puabi of Ur. A copper tube was encased in lapis-lazuli. A silver tube with gold and lapis rings measured one centimeter (3/8 inch) in diameter and 93 centimeters (3 feet) in length.

The Sumerians used their grain harvest to produce over 300 types of bread, generically called NINDA. After the grain was ground into flour using the saddle quern, flat breads were made simply from a mixture of grain flour and water, plus possibly a little salt. The dough was placed on a hot stone or inside an Arabic tannur oven. Sometimes the dough would be placed directly in the ashes to bake. These flat breads could not be stored so they would be eaten immediately and washed down with beer. Leavened beer breads were also baked, apparently using yeast obtained from brewed beer.

**Figure 7.3.** Drawing of Sumerians drinking beer from vats through long tubes. (From Singer *et al.* 1954)

Overall, crop and food production in Sumeria required unrelenting work with no respite throughout the year from the demands of growing crops. Crops matured in March and April and required immediate harvest, followed by threshing. The completion of each harvest meant attention must be turned to the next crop. That required hoeing the fields as soon as weeds began to emerge. Threshing and hoeing probably occupied most of the first part of the summer. These jobs were followed with maintenance work on the canals and dikes to remove accumulated silt and repair any damage to the canals and dikes. It was essential to have the irrigation system ready to distribute water before sowing. In November, the next crops were sown. The months while the crop was growing were occupied with continual attention to distributing irrigation water and removing weeds from the fields. Of course, the daily chores of grinding malt and grain, brewing beer, and baking bread never ended. Only a few in this society had the luxury of pursuing activities other than growing crops or preparing food and drink.

While the Sumerian society persisted for more than 1000 years, the eventual destruction of this society was an inherent consequence of the cropping system on which they depended. Silt and organic matter carried by the Euphrates River were key to maintaining the soil and fertility of the Mesopotamia Valley. Water and nitrogen for a new crop were replenished each year by irrigation water flowing onto the fields. Unfortunately, river water carried mineral salts to the fields and irrigation of fields over centuries resulted in an accumulation of salt in the soil. Wheat is sensitive to salt, and the soil of the fields of Mesopotamia eventually contained too much salt to grow wheat. Barley is more tolerant of salt, so as fields became saline wheat yields decreased and grain harvests shifted to increasing proportions of barley. Eventually barley would be the only crop that would grow in the most saline fields. The Sumerian reputation for high beer consumption was surely aided by the fact that barley – more suitable for beer than bread – persisted as salts accumulated in the soil of their fields.

In addition to salinity, flooding was a more obvious danger to Sumerians. The slow flow of the Euphrates River in the Mesopotamia Valley allowed silt in the water to settle, forming natural silt dikes and raising the elevation of the river channel. It is likely the silt load of the Euphrates at the Taurus Mountains increased throughout the Sumerian period because the hillsides were being cleared of vegetation by grazing and even some crop production. The deforestation of these mountains would have resulted in increased runoffs from rains and more soil erosion. Flood was an ever present danger because the channel of the Euphrates River was elevated even higher than some of the crop fields. A river higher than the adjacent land made the land extremely vulnerable to devastating floods. Early spring when the water from the winter rains in the Taurus Mountains reached the Mesopotamia Valley must have been a terrifying time as the river rose. The fields were full of ripening grain. Would the rising river over-top the dikes? Would seepage through the natural silt dikes of the river cause collapse of the dikes and a devastating flood on the fields? Not only would a flood jeopardize the existing crop, it could fill the irrigation canals with silt and sweep away the dikes around fields used to contain irrigation water. Recovery from a major flood must have taken years.

Flood devastation was an important part of Sumerian folklore or oral history. It is alluded to in the Sumerian epic of Gilgamesh. Indeed, it has been suggested that the origin of the story of Noah and his Ark came from these experiences with flooding. A major flood resulting from the combination of high winter rainfall in the Taurus Mountains and heavy spring rains in the nearby Zagros Mountains would flood the "whole world" encompassed in the Mesopotamia Valley. The story of Noah could be an account of a merchant living on the river where his family traded in grain, beer, and livestock. Caught in a flood, Noah and his Ark could have been swept into the Persian Gulf where he and his family, with the animals on board, floated until reaching a distant shore. Assuming the boat was well stocked with grain and beer, survival for "40 days" would have been possible.

As devastating as floods would have been, aggravation of the salinity problem as a result of the elevated height of the Euphrates River would prove to be even more devastating and insidious in undermining the survival of the Sumerian society. Salts continue to accumulate in irrigated fields unless the salt is somehow flushed from the soil. Periodic floods may have helped to remove some of the salts, but sustained removal of salts was blocked by the high river channel. There was nowhere for the saline water in the soil to move when the river channel was higher than the fields. Salt was trapped under the fields. In addition, the elevation of the river resulted in a hydrostatic pressure preventing the flow of salt in the soil away from the fields. As a result, saline water was trapped in the fields, and the hydrostatic pressure forced the saline water to move up into the crop rooting zone. Ultimately, salt concentration built up in the fields until crops could no longer be grown. The salt marshes in modern southern Iraq are a remnant of the salt accumulated on Sumerian fields.

To survive, the Sumerians had no choice but to move to new lands that had not yet been heavily irrigated. In the Mesopotamia Valley, Sumerian migration to new lands meant moving further north along the river. Eventually, the Sumerians came into conflict with the Akkadians entrenched in the north. While there are always a number of factors contributing to a transition in power, one of the points of conflict and weakening of the Sumerian societies almost surely was salinization of the Sumerian crop lands. The Akkadians ascended to power in the Mesopotamia Valley in 2334 BCE.

The Akkadian empire was the first centrally organized empire in the world and encompassed much of the Fertile Crescent. The Akkadian empire lasted little more than one century, until approximately 2220 BCE. The fall of the Akkadian empire coincided with a long-lasting drought in the Middle East. Based on sediment cores taken from the bed of the Gulf of Oman, a major drought developed in the Middle East by about 2200 BCE. Refugees from outside the Mesopotamia Valley would have been forced to flee to the productive valley to avoid starvation. In addition, the drought conditions would have resulted in less flow in the Euphrates River, making irrigation more difficult. The increased population and the challenges in maintaining a viable cropping system may have been important contributors to the downfall of the Akkadian Empire.

## Sources

Bottero J (2004) *The Oldest Cuisine in the World; Cooking in Mesopotamia.* Translated by T Lavender Fagan. The University of Chicago Press. Chicago.

Burke J (1978) *Connections.* Little, Brown and Company, Boston.

Curtis RI (2001) *Ancient Food Technology.* Brill, Leiden.

Dale T, VG Carter (1955) *Topsoil and Civilization.* University of Oklahoma Press, Norman, OK.

Harlan JR (1992) *Crops & Man.* American Society of Agronomy, Madison, WI.

Hillel DJ (1991) *Out of the Earth. Civilization and the Life of the Soil.* The Free Press, NY.

Jacobsen T, RM Adams (1958) Salt and silt in ancient Mesopotamian agriculture. *Science* 128:1251-1258.

Kiple KF, CO Kriemhild (2000) *The Cambridge World History of Food, Volume Two*. Cambridge University Press, Cambridge, UK.

Molleson T (1989) Seed preparation in the Mesolithic: the osteological evidence. *Antiquity* 63:356-362.

Singer C, EJ Holmyard, AR Hall (1954) *A History of Technology*. Oxford Press, London, UK.

Smitha F (n.d.) *Map of Ancient Mesopotamia, to 2500 BCE*. [Online] Available at: http://www.fsmitha.com/h1/map01mes.htm [Accessed 26 January 2010].

Tannahill R (1973) *Food in History*. Penguin Group, UK.

Trigger BC (2003) *Understanding Early Civilizations: A Comparative Study*. Cambridge University Press, Cambridge, UK.

# 8

# Egyptians

# ~3000 to 1070 BCE

*Hail to you Hapy, Sprung from earth, Come to nourish Egypt…*
*Food provider, bounty maker, Who creates all that is good!…*
*Conqueror of the Two Lands, He fills the stores,*
*Makes bulge the barns, Gives bounty to the poor.*

From a Middle Kingdom hymn
translated by Lichtheim

Hapy, the god who personified the Nile River, was the giver of all good things in Egypt. Indeed, the word Nile is derived through a series of translations of Hapy. Living in the middle of a desert, the Egyptians revered the water of the Nile River. Harsh desert isolated and even protected Egypt from invaders, and the Nile provided the water needed by both the people and their crops. On the banks of the Nile the Egyptians were secure and could develop strong societies. Growing of crops in the Nile Valley may have started as early as 5000 BCE, likely as a result of contact with people originating in Mesopotamia. Certainly, wheat and barley were among the first crops grown in Egypt.

In about 3000 BCE, King Narmer unified the communities along the Nile north of the first cataract (Fig. 8.1) into a single country. From this beginning evolved a society with a unique cultural identity that persisted for two millennia. The Egyptians developed a very complex society with the pharaoh at the pinnacle assuming the position of a living god. The pharaoh was surrounded by a court including priests and artisans. The pharaoh owned all the land and claimed the bounty produced on these lands. The elite class controlled the production and distribution of much of the food supply and imposed taxation which kept the granaries of the pharaoh full. The elite class was well fed on the cereals produced in Egypt as well as exotic food imports and specialized local foods that involved intensive labor. The vast majority of the people, however, were relegated to diets of mainly beer and bread made from wheat and barley grown on the fields surrounding the Nile River.

It is speculated that the origins of the wheat and barley crops grown in Egypt were originally seed imports from the Mesopotamia Valley. Unlike the challenges of growing cereals in the shadow of the Euphrates River, the Nile was truly a beneficent river with all the attributes needed for sustained production of wheat and barley. It is not surprising that Hapy was second only to Ra, the sun god, in the pantheon of Egyptian gods. Thanks to the unique characteristics of the Nile River, cereals could be grown with comparative ease and yields were usually bountiful.

Rather than fear flooding of the Nile, the Egyptians welcomed the flooding that restored water to the soil on which crops were sown. Indeed, the flood of the Nile occurred at the most advantageous time of year to fully water the fields just before the cereals were sown. Wheat and barley grow very poorly in the heat of the summer months, especially in an environment as hot and dry as Egypt. Therefore, the optimum time for sowing these species is late in the fall so that the plants would grow during the winter months and mature in the spring. The good fortune of the Egyptians was that the peak of the annual Nile River flood occurred in early October (in the modern calendar), covering the fields so that they were fully recharged with water after the long dry summer. The land emerged from the flood waters in November, the optimum time to sow winter cereal crops.

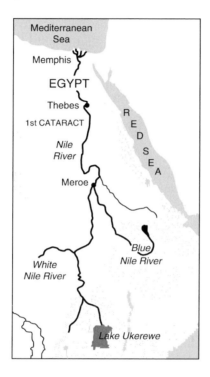

**Figure. 8.1.** Map of Nile River system originating with the White and Blue Nile Rivers in central east Africa. Ancient Egypt occupied the Nile River Valley only north of the first cataract. (Redrawn from Smitha (n.d.).)

The Egyptians explained their good fortune in the timing of the Nile River flood by mythology. The Nile River is actually fed by monsoonal rains that fall well to the south of Egypt in central Africa during the summer months. The Nile River originates at the confluence of the Blue Nile and White Nile Rivers (Fig. 8.1), and both rivers are fed during the summer by torrential summer rainstorms. The flood waters reached ancient Egypt beginning in late summer with the peak usually occurring in October. In November the flood water would recede leaving the soil fully watered just in time for sowing of the new crop.

Additionally, the Nile River provided much more than water. The flood waters of the Nile carried essential nutrients for the new crop. The Blue Nile originates in Ethiopia's east highlands, a very rugged terrain of exposed soil and rock. Due to the absence of rainfall in the winter months and cool temperatures, there is little plant growth to protect the hillsides. Therefore, once the summer storms begin, the hillsides are readily eroded, scouring away soil and rock. Fortunately for Egypt, the eroded rock contained the mineral nutrients phosphorus and potassium that are essential for good plant growth. Annual flooding of the Nile in Egypt deposited these minerals for use by the new crops.

The White Nile River originates farther south and to the west of the Blue Nile (Fig. 8.1). The water for the White Nile comes from the monsoonal rains that fall on the savannas of central Africa. This water flows into large, swampy wetlands located in modern southern Sudan. Vegetation grows abundantly in these wetlands nearly year round. The summer rains cause the water levels to rise in the wetlands, loosening new and decaying plant material. This freed plant material is the organic matter carried into the White Nile, and eventually into the Nile. The organic matter, when deposited on the wet fields in Egypt, decays to provide the nitrogen that is essential to sustain growth of cereal grains on the Nile flood plain.

The flood cycle of the Nile River also was a huge benefit in weed control in growing wheat and barley. Unlike Mesopotamia, where the fields needed to be continually plowed in the months preceding sowing of the crop to control weeds, the flood cycle of the Nile River provided the necessary weed control before sowing a new crop. During the summer months the Nile was at low flow and the river channel was below the elevation of the fields. Hence, there was no hydraulic pressure causing water to flow under the soil into the fields. The soil would dry during these summer months making it difficult for unwanted seeds to survive and weeds to grow. Weeds that did establish would be suffocated when covered with the flood waters. Therefore, the annual flood of the Nile River would have effectively cleared the fields of weeds. Virtually no field preparation was needed before sowing the fields; the Nile obligingly watered the fields, added a new dose of fertilizer, and removed the weeds. Crop seeds could be simply broadcast on top of the one-millimeter (1/25 inch) thick layer of new, muddy, nutrient-rich soil.

Potential weed seeds would also be carried to the fields from the White Nile River. However, it is likely that these imported seeds from plant species native to the hot climate of central Africa would not grow well during the cool, winter months of Egypt to compete with the wheat and barley plants. In addition, it may be that the Egyptians developed a management scheme to further control weeds without having to resort to a massive effort to manually remove weeds. We propose that animals were used to eat some of the weed plants. It has been reported that animals were herded on to the muddy fields after crop seeds were broadcast. One interpretation has been that the animals helped to bury the seeds by walking in the fields. However, seeds would have readily germinated in the muddy conditions in the fields at this time of year without burying. Further, the hooves of animals would have likely pushed the seeds too deeply into the muddy soil and resulted in poor plant emergence. Alternately, we speculate that the reports of farmers putting animals into the fields may have referred to allowing the animals to graze the fields once the cereal crops had emerged. The dual use of wheat as a forage early in the season and as a grain crop later is practiced today in many parts of the world, including the U.S. During the early stages of wheat and barley growth, their leaves develop above the soil but their growing tips remain protected from grazing below the soil surface. There is little loss in cereal yield by allowing animals to graze the plants in the first few weeks of growth. On the other hand, a major advantage in allowing the animals on these fields is that they would likely also eat any emerging weed plants. Since the growing tips of any non-grass weeds would emerge above the soil, the removal of their growing tips would have substantially inhibited any further growth of these weed plants. This use of livestock grazing would have been very helpful in weed control. Additionally, animals benefited from their first fresh forage after a long, dry summer.

Unlike the Sumerians, the Egyptians had little need for elaborate irrigation schemes. In most years, the flood of the Nile River was sufficient to fully fill the soil with water. The roots of these cereal plants readily extract water from 1.5 meter (nearly 5 feet) deep into the soil. This volume of soil would have stored at least 200 mm (8 inches) of water for use by the subsequent wheat and barley crops. Evaporation of water from these wheat and barley crops during the cool winter months would have been substantially less than the evaporation in the summer. Generally, these winter crops would have required no more than the 200 mm of water stored in the soil. Therefore, after the Nile River flood filled the soil profile with water, there was sufficient water in the soil for the entire crop growth cycle, and there would be no need to irrigate these winter wheat and barley crops.

In some years, however, the Nile River would not have carried enough water to flood all fields. In years when the monsoonal rains resulted in less water, the Nile flood likely covered only the lowest-lying fields. Higher elevation fields would either not be sown or have poor yields. Therefore, the loss of productive lands in the years of low flood could have resulted in food shortages even in the Nile River valley. Moses' prediction of seven years of famine was a weather forecast about the monsoons at the headwaters of the Nile River.

In a normal flood year, sowing of the wheat and barley crops could begin on higher elevation fields as soon as the muddy soil surface emerged from the water. Hence, a natural staggering of the sowing dates, and ultimately harvesting dates, would allow the demands for labor at the beginning and the end of the growing season to be spread over several weeks.

There are historic references to Egyptian farmers irrigating fields using a system of canals and a shadoof to lift water. The shadoof consisted of a long pole suspended on an upright frame. A bucket was placed on the long end of the pole and a counter weight on the short end. The bucket could be dipped into a water source and lifted to put water in the irrigation canal. Since there was no need to irrigate wheat and barley, the shadoof and irrigation system were likely used at other times of the year. If warm-season crops were grown in the summer months to augment the Egyptian diet, it would surely have been necessary to irrigate these gardens. Hence, irrigation was a feature of Egyptian agriculture, but it was likely used only on small garden plots on which summer specialty crops were grown.

Wheat and barley crops would have matured in March, and harvest would occur in April. The grain heads were cut from the plants with sickles and taken for threshing. Threshing was accomplished by beating the grains from the grain heads. One technique used by the Egyptians was to pound the grain heads in a mortar, likely made of limestone, using meter-long wooden pestles. Two men worked together to pound up and down with the pestle in a tall mortar, or round-and-round in a shallow pestle. The separated grain could then be stored in mud-brick silos. Grain storage was facilitated by the very low atmospheric humidity of the Egyptian environment. An interesting possibility is that storage in these silos might have fostered *Streptomycetes* microbes to grow under some conditions. These microbes produce antibiotic tetracyclines that may have provided the consumers of beer made from these grains some degree of immunity against infection.

The grains would be removed from the storage silos as needed. Grains were ground into flour, usually by women using a saddle quern (see Fig. 4.1). Separation of the hulls from the ground flour was accomplished by putting the ground material through wicker sieves made of reed, rush, or palm. The coarse mesh of the sieve separated out only the largest particles, so the process would be repeated if a higher-quality flour was desired. The ground flour was used to make bread. The oldest and simplest method was to prepare unleavened bread by forming the dough into a pancake and baking it directly in the ashes of a fire. The dough could also be cooked on flat stones heated in the fire.

The Egyptians appear to have eaten mainly leavened bread. In this case, the baker worked with hands – or feet – to knead a mixture of the flour and water into dough. Fermentation would be initiated in the wet dough within hours as airborne wild yeast produced sour dough. A variant in bread making was to mix the dough with beer. In this case, the rise of the dough would have been more spectacular than that achieved with a little airborne yeast. The risen bread was then formed into the desired shape by hand or in molds of great variety. Early Egyptian molds were made rough on the exterior and coated on the interior with a thin fine-grained clay. They were rather thick containers measuring 13 to 23 centimeters (5 to 9 inches) in exterior height, 18 to 25 centimeters (7 to 10 inches) in diameter, and 7 to 14 centimeters (3 to 6 inches) in interior depth. These molds were heavy, weighing about 3.3 to 6.5 kilograms (7.3 to 14.3 pounds). Since these molds have been recovered from the houses of commoners, they were probably the usual container for daily baking. Molds with dough were placed in a variety of ovens for baking.

Substantial quantities of beer were also produced from the stored grains. Grains would be malted, ground to a fine texture, and the resulting flour would be baked into loaves. The dough could be flavored with juice from dates or pomegranates to add flavor to the beer. The beer loaves would be only partially baked. These partially baked loaves would be broken up and mixed into water, and then allowed to ferment. The fermented liquid was separated using a mesh screen. Some of the bread mash would also be forced through the screen. The mixture would be placed in jars and allowed to ferment further. This beer was consumed by Egyptian men, women, and children on a daily basis more as a part of the food diet than a beverage. Since the beer was consumed fresh and contained particles of the bread, it was likely to have been quite nutritious. Such a beer would have been an excellent source of protein and B vitamins.

Production of beer and bread was not confined to the home. Large quantities were produced in buildings constructed for brewing on one side and baking on the other. Figure 8.2 is a drawing of one of these buildings, which shows the numerous tasks required in producing both beer and bread. Half-baked bread loaves produced on one side were used in brewing, and beer produced on the other side was used as an ingredient in making bread dough. The building architecture itself is a testament to the connection between these two products.

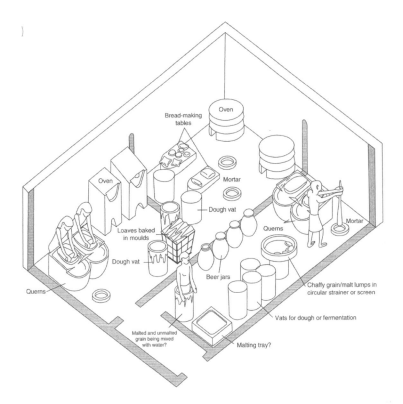

**Figure 8.2.** Drawing of Egyptian building for baking (on the left) and brewing (on the right) from the Twelfth-Dynasty Tomb of Meket-re at Thebes. (From Kemp 2006.)

The work cycle through the year was less demanding in Egypt than in many societies. While women worked year round to bake bread and brew beer, a large labor force of men would only be required in the fields from sowing in November through harvesting in April. Consequently a large labor force of men was available each year for nearly six months, May to October, for other activities such as constructing the many impressive

Egyptian monuments. Since the lands were flooded during the later part of this non-cropping period, the high waters would have likely facilitated floating large stones on rafts to the construction site of the various monuments. The laborers were paid in the bread and beer produced from the grains claimed by the pharaoh. It appears that a key factor in allowing Egyptians to build some of the largest monuments ever constructed in the ancient world was the unique characteristic of the Nile River flood which allowed relatively easy and abundant food production.

---

**Box 8.1.** Pregnancy Test
Bakers apparently had an important task beyond that of baking bread. Egyptian women used wheat grain obtained from bakers for a pregnancy test. Women would urinate on these wheat seeds. Seeds were thought to germinate in sympathy for a pregnant woman and failure to germinate indicated lack of fertility.

---

Why did the Egyptians not suffer salinity problems resulting from a dependence on river water to irrigate their fields? Again, the nature of the Nile River flood avoided such salinity problems. The large annual surge of flood water caused silt deposited in the channel or banks to be swept away. Therefore, the elevation of the Nile did not rise as did the Euphrates, and after the annual flood the Nile River receded back to a relatively low channel. Consequently, there was no hydraulic pressure from the river blocking seepage back to the river of the water and the salt it contained. The water receding from the fields effectively flushed any accumulated salt from the soil. There are no accounts of salinity problems of fields adjacent to the Nile River in antiquity.

The zenith of Egypt's power and influence was initiated in about 1550 BCE with a reunification of Egypt and the initiation of the New Kingdom. The large and dependable grain supplies were the underpinning for a period of expansion, and much of the Middle East came under the control of the Egyptians. However, the imperial expansions eventually backfired as Egyptian organization and technology were adopted by those they conquered. Rebellions in these conquered lands weakened the Egyptian dynasty, and the New Kingdom ended in 1070 BCE.

Ultimately, Egypt came under the domination of Athens and Rome. As discussed in Chapter 12, during the Golden Ages of both of these societies neither was capable of meeting the demand for wheat by domestic production. Hence, to feed their people and sustain power it was essential to control lands where grain production was abundant. Egypt was, of course, a lucrative and convenient target with massive grain production. In succession, Athens and then Rome conquered and controlled grain export from the Nile Valley. The pharaonic era of Egypt was over and the bounty of the Nile Valley continued to be exploited in grain production, but now to support the Golden Ages of foreign powers.

## Sources

Allen SL (2002) *In the Devil's Garden: A Sinful History of Forbidden Food.* The Ballantine Publishing Group, NY.

Bottero J (2004) *The Oldest Cuisine in the World: Cooking in Mesopotamia.* Translated by T Lavender Fagan. The University of Chicago Press, Chicago.

Curtis RI (2001) *Ancient Food Technology.* Brill, Leiden.

Dale T, VG Carter (1955) *Topsoil and Civilization.* University of Oklahoma Press, Norman, OK.

Hillel DJ (1991) *Out of the Earth. Civilization and the Life of the Soil.* The Free Press, NY.

Iliffe J (1995) *Africans: The History of a Continent.* Cambridge University Press, Cambridge, UK.

Kemp BJ (2006) *Ancient Egypt: Anatomy of a Civilization*. Routledge, UK.

Smitha F (n.d.) *Map of Africa, 2500 to 1500 BCE*. [Online] Available at: http://www.fsmitha.com/h1/map02af.htm [Accessed 26 January 2010].

Tannahill R (1973) *Food in History*. Penguin Group, UK.

Trigger BG (2003) *Understanding Early Civilizations: A Comparative Study*. Cambridge University Press, Cambridge, UK.

# 9

# Chinese

# 221 BCE to 220 CE

*The most precious things are not jade and pearls but the five grains.*

Shennong
Mythical Emperor
of the Five Grains

Chinese society originated in the northern areas of modern China. Similar to the Sumerians and the Egyptians, the Chinese settled on a relatively flat flood plain to grow their crops. The Yellow River (Huang He) and its tributaries formed the flood plain on which the Chinese could grow their crops (Fig. 9.1). The Yellow River originates in the highlands in the north and west, which were created over the millennia by soil blowing in from the arid central plains of Asia. Since the loess soil from this region has a yellowish color, the river was called the Yellow River. The accumulated loess soil (wind-blown) has a silt texture and is very deep, and under most circumstances would be ideal for plant growth. However, the highland region is arid and the altitude is high with cold temperatures nearly year round. Plant growth is sparse on these loess deposits and consequently even modest rains or melting of winter snowfall results in a large surge of water and soil runoff. The eroded runoff from these highlands fills the Yellow River with a silt content among the highest of any in the world.

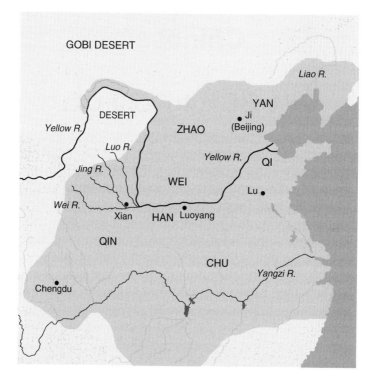

**Figure 9.1.** Map of Han Dynasty showing the regional states before unification by Qin Shi Huangdi. Xian was the capital city established by Qin Shi Huangdi. (Redrawn from Smitha (n.d.).)

Plants were domesticated on the Yellow River flood plain, and growing of crops became the fundamental basis for subsistence of these early peoples. However, the Chinese had a very different relationship with the Yellow River than did the people depending on river waters in the Mesopotamia and Nile Valleys. Instead of being a giver of life, the Yellow River was viewed as a constant threat of flooding and widespread devastation. The very high silt content of the river caused the river channel and river banks to rise and loom over the surrounding countryside. A breach of these weak silt banks would unleash the river to inundate all surrounding fields and villages. One estimate indicates there were sixty-eight major floods during the Han Dynasty, or about one major flood every six years. The Yellow River was no gift of a deity – it was labeled the River of Sorrow.

Until the middle of the third century BCE, the Yellow River region was fractured and controlled by local warlords (Fig. 9.1). However, China was soon to be consolidated under a single emperor. In an effort to alleviate some of the constant threat of flooding, the warlord of the province of Qin (pronounced Chěn, a term later applied to the whole country of China) commissioned engineers in the middle of the third century BCE to control the flood threat and potentially provide irrigation water for crops. In 256 BCE, engineer Li Bing began construction of channels and banks that harnessed the Min River. Ten years later Sheng Guo, another engineer, constructed a canal between the Jing River and the Luo River that was 150 km long and opened 40,000 hectares (100,000 acres) for growing crops. These water projects resulted in a significant increase in crop productivity and greatly increased the wealth in this region. The leader who ultimately benefited from this new agricultural productivity and wealth was Qin Shi Huangdi, who ascended to the Qin throne in 247 BCE as a 13-year-old boy.

As a young man Qin Shi Huangdi consolidated his power by taxing the growing crop productivity of his farmers. He used this wealth to organize a large army to attack and conquer a succession of neighboring states. By 221 BCE, he had consolidated nearly all of the northern regions of modern China into a single very large realm. He established a powerful central government and ruled the country with brutality. Roads were constructed, laws were standardized, and a common language was introduced across the country. Qin Shi Huangdi also ordered the beginning of construction for the Great Wall to protect China from barbarian invaders from the north. Farmers were conscripted as forced labor to build the Wall, even though the loss of their agricultural labor resulted in starvation. Also, Qin Shi Huangdi is remembered for his huge burial complex near what became the capital of the Han Dynasty near Chang'an (Xian, Shaanxi Province), which included legions of life-sized terra cotta soldiers to support him in sustaining his military power in the afterlife.

Qin Shi Huangdi died in 210 BCE, only eleven years after the consolidation of China. Following a brief period of civil warfare, Liu became emperor of China in 206 BCE and established the Han Dynasty that ruled China for more than 400 years. The Han dynasty is sometimes referred to as the Golden Age of China; arts and literature flourished during this period. Taxation of the farmers was often fairly moderate, encouraging farmers to high productivity. High agricultural productivity was sustained through much of the Han Dynasty and resulted in a fairly peaceful period in Chinese history.

The Chinese faced unique challenges in growing crops to achieve high productivity on the flood plains of the Yellow River and its tributaries. While the friable loess soils of the Yellow River flood plain were fairly easily worked, the river offered few other benefits to the Chinese. Irrigation was a dangerous proposition. Breaching the natural dikes of the Yellow River opened a weak point in the loess banks and increased the risk of flood. Fortunately, the benefits of the temperate climate of northern China, which often provided adequate summer rainfall for growing crops, meant the Chinese were not dependent on irrigation water. The primary focus of many river and canal projects was directed at controlling floods, or facilitating transportation, and allowing drainage of swampy lands.

The Yellow River also failed as a source of nutrients to fertilize the fields, in contrast to what had occurred in the Euphrates and Nile valleys. Although the waters of the Yellow River were filled with huge quantities of eroded silt, the poor plant growth in the eroded highlands meant that the water carried very little of the organic matter which would have been a source of nitrogen for the growing crops. Failure of the Yellow River to provide a steady influx of nutrients, particularly nitrogen, forced the Chinese into a number of unique solutions to obtain the required nitrogen for their fields. All sources of nitrogen were exploited for use as fertilizer. One immediate solution was to rely on crops that were less sensitive to nitrogen deficiency than other plant species. At the time of the Qin and Han Dynasties, the principal staple crops were spiked millet (*Setaria italica*) and short-season, common panicle millet (*Panicum miliaceum*). A major advantage of millets as the main staple crop is that they grow well on less nitrogen and water than required by wheat and barley.

The smaller requirement for nitrogen by millets is a result of the additional biochemical sequence in their photosynthetic pathway that captures carbon dioxide from the atmosphere without a heavy investment in nitrogen. This extra sequence has a much higher affinity for carbon dioxide than the basic enzymes of photosynthesis. Hence, millets do not need a large amount of the basic enzymes, and their leaf nitrogen content is lower than crops without the extra biochemical sequence. Millets can have a good growth rate with a much lower nitrogen uptake. Further, millets have a short growing season, so the plants remain fairly small and do not need to accumulate large amounts of nutrients. Overall, the use of millets as the basic staple grain fit well with the low fertility conditions that existed in China.

Nevertheless, some nitrogen must be made available even to the millet crop, and the Chinese gave considerable attention to cycling animal and human manure back to the soil. The primary animals raised on the farms were poultry and swine, which have a digestive system that results in more excretion of nitrogen than do animals with ruminant digestion, such as cattle. Of course, humans also have non-ruminant digestion, so human waste was a valuable addition to the soils – the origin of the use of "night soil" by the Chinese.

Other sources of nitrogen for the fields were crops that incorporate atmospheric nitrogen into plant components by symbiotic nitrogen fixation. One important crop species that has the ability to support symbiotic nitrogen fixation is soybean. Soybean can be grown without the addition of any nitrogen fertilizer to the soil. If the seed yields are not high and a substantial portion of the nitrogen is left in the leaves and stem, the vegetative parts of the plants can be incorporated into the soil as organic matter to contribute nitrogen to subsequent crops. (In modern, high-yielding soybean systems, most of the nitrogen ends up in the seed and is consequently harvested and removed from the fields.)

---

**Box 9.1.** Rice

Notably absent from the list of major crops in the Han period was rice. Rice was surely grown and consumed in southern China on the plains of the Yangtze River where the temperatures were mild and the growing season long. Most importantly, the southern wetlands of the Yangtze River allowed the establishment of paddies where rice could be grown for most of the season immersed in water. The production of rice under flooded conditions was a very effective way to kill most weeds that would naturally compete with rice. However, in northern China during the Han Dynasty the amount of available water and an ability to control the level of water in paddy fields was insufficient for extensive production of rice. The elite of the Han Dynasty likely had rice as part of their diet, but not the common farmers. (It appears that some confusion about farm production originates from an early mistranslation of the word "grain" to mean "rice".)

Rice, of course, is another grass species and has many of the same attributes as wheat, barley, and other grasses. The unique feature of rice that allows it to be grown in flooded conditions is air channels running from its stems through the roots (aerenchyma). Oxygen can diffuse through these air channels to the roots allowing the plant to flourish in flooded paddy fields. Another advantage in growing rice is that nitrogen-fixing cyanobacteria form symbiosis with blue-green algae growing in the paddy water. These organisms add small but important amounts of nitrogen to the crop system. The added nitrogen was sufficient for low-yield rice production.

Rice grain is abundant in starch and can be readily fermented to produce beer. Rice has an agreeable flavor that when cooked makes a very satisfying food by itself. Its high starch content provides the calories needed for the intense labor of tending the paddy fields. However, the low protein content does not give high nutritional quality. The advent of polished rice – in which the outer layers of the seed were removed – resulted in a loss of a major source of protein including the amino acid thiamine. As a result, consumers of polished rice were vulnerable to the disease beriberi as a result of thiamine deficiency.

**Box 9.2.** Soybean

Soybean is about the only major crop both in antiquity and in modern times that is not a grass. The unusual importance of soybean is a result of its ability to acquire nitrogen from the atmosphere and the high oil and protein content of its seed. Specific bacteria in the soil can infect the roots of soybean, triggering a proliferation of cell growth that results in a sphere from 1 to 6 mm (1/25 to 1/4 inches) in diameter, called a nodule. Inside the nodule the bacteria population explodes. Instead of causing disease, however, the bacteria have the special ability to incorporate atmospheric gas into organic compounds. The organic compounds are released to the soybean plant, which transports them to the top of the plant for growth of the plant and seeds. The symbiotic arrangement between the bacteria and soybean plant is exploited in agriculture to alleviate the need to provide external sources of nitrogen.

Soybean produces a seed of high oil and protein content, in contrast to cereal grains. Soybean oil, which is quite desirable in cooking, is obtained by compression. Soymilk is made by soaking the seeds and then grinding in hot water. During the Han Dynasty tofu production originated. Tofu was obtained by adding a coagulant, commonly gypsum, to soymilk. The resulting product provided 320 kilocalories per 100 grams containing 8% protein and 5% fat. Tofu was used in many ways, including an addition to stir fry and to the gruel prepared from millets. Also, soybean seeds that are not quite mature can be consumed directly after boiling in water, producing the Japanese dish edamame.

In addition to millets and soybean, wheat was a major component of the Chinese cropping system. Wheat grows best under cool temperatures, so in the temperate climate of northern China wheat would have been sown in the fall and harvested the following summer. Wheat was already grown in the Han period as an alternate-season crop in addition to the summer crops of millet and soybean. In the early Han Dynasty, wheat was not a particularly desired food, but it offered a grain yield at the end of winter as an emergency food in case a severe drought brought crop failure in the summer. Writings from this period discuss various alternatives in the cropping rotations among the three or more crops.

The common crop rotation scheme used by the Chinese seems likely to have included short-season millet, wheat, soybean, and spiked millet. A three-year rotation was likely practiced in each field (Fig. 9.2). In the first year, a short-season millet would mature in late August/early September of the modern calendar, and the wheat crop could have been sown immediately after millet harvest. After the harvest of the wheat in the spring of the second year, the soybean crop would have been sown. Soybean seeds were likely harvested during both a fresh stage and at dry maturity. In the third year, the soil would have benefited from the soybean crop, and the favored crop of spiked millet would have been sown. Therefore, this rotation allowed four crops in three years, and it gave the desired balance of two millet crops and one each of wheat and soybean crops. Such complexity in crop rotations was not matched in the West for about another 1000 years.

| Field | Summer | Winter | Summer | Winter | Summer | Winter |
|-------|--------|--------|--------|--------|--------|--------|
| #1 | Short-season Millet | Wheat | Soybean | – | Spiked Millet | – |
| #2 | Spiked Millet | – | Short-season Millet | Wheat | Soybean | – |
| #3 | Soybean | – | Spiked Millet | – | Short-season Millet | Wheat |

**Figure 9.2.** Crop rotation sequence commonly used during the Han Dynasty to grow four crops in three years in three fields.

During the Han period, families were assigned farms that averaged only about three hectares (eight acres). To feed the family and pay the crop taxes, productivity had to be maximized. The land would be subdivided into small fields of about 0.1 hectares (0.2 acres) to allow a staggering of the rotation sequence so that each stage of the rotation was in place in each year. Several small fields at the same stage in the rotation cycle could be staggered in the specific dates of sowing and harvesting. This scheme allowed the workload to be distributed over a greater range of dates among the several small fields growing the same crop, and thereby the time demands in the field would be spread over a longer period. Also, staggered dates in the growth of each crop among several fields offer some protection against environmental stresses to which the crop may be vulnerable at a specific growth stage. Using the rotation described above, work in the fields would have been nearly continuous. In the early spring, short-season millet would have been sown in one field first. Next, in a second field the spiked millet would have been sown. The next task would have been to harvest the wheat crop, which had been sown on a third field the previous fall and then immediately re-sow that field with soybean. The short-season millet would be harvested in late summer and the field re-sown with wheat. The harvests of the spiked millet and soybean would have likely followed as the final harvests of the year. Weeding and manure application of the fields would have continued year round.

The Chinese in the time of the Han Dynasty also developed a unique tillage system in their fields, called tai-t'ien, to accommodate the crop rotation system. Each small field was formed into a series of trenches and ridges with a height difference of about 25 centimeters (10 inches). A v-shaped plow would be pulled through the field to form these trenches and ridges. Murals in the burial tombs of the rich show oxen pulling iron plows, which may have been the case on large manors. However, for the ordinary small farm it is more likely that the plow to form the trenches would have been repeatedly pulled through the field by humans. Once the trenches were formed, seeds were sown in the bottom of the trench. There would have been several advantages to this ridge system in contrast to broadcasting seed on the soil surface. The trenches protected the young seedlings from early spring's cold, strong winds which could damage leaves and dehydrate plants, and also from a late frost. Therefore, the millet crop could be sown early in the spring. Finally, trenches helped in weeding. As young plants gained stature, soil from the ridges could be pushed into the trench to cover the base of the crop plant and smother any emerging weeds. As the season progressed, the trench would eventually be filled with soil, or even hilled up against the base of the crop plant. Therefore, the possibility of damaging a crop plant by trying to hoe close to the plant was minimized; the soil only needed to be pushed to cover the weeds. Numerous iron hoes from the Han period have been identified that would have been effective for use in the tai-t'ien system.

Once the crops matured, the grain heads would be removed from the plants with iron sickles and gathered for threshing. An innovation of the Han period was pedal-operated "hammers" that repeatedly pounded the grain heads (Fig. 9.3). Basically, the system consisted of a long pole balanced on a fulcrum with the operator using his feet to raise and lower the hammer onto the grain. Wind or hand-cranked fans were used to blow the chaff from the grain kernels. The grain was then stored until needed. For the early part of the Han era, the grain was ground into flour using a mortar and pestle, some were made of bronze.

**Figure 9.3.** Chinese grain threshing. Pedal-operated hammers on left and hand-cranked fan for chaff and grain separation on right. A storage building is in the background. (From Lowe 1968)

Millets were very much the crops preferred by Chinese because they gave a desirable, nutty, somewhat-sweet flavor when fermented to make beer or consumed as food. While translations often refer to the Chinese drinking rice wine, this is almost surely a mistranslation from Han period writings; rice was not grown in quantity in northern China. Rather, the favored drink of the Han was beer fermented from millets. Herbs were often added in the brewing, supposedly to boost the medicinal quality of the beer. Herb-flavored millet beer was consumed in quantity by the Chinese. In fact, brewed millet was seemingly drunk in sufficiently large quantities that laws were enacted to suppress alcohol excesses.

Millets were also used directly for food. One of the easiest food preparations was simply to cook the grain in water, and within a few minutes a fluffy porridge would result. However, the food consumed by all Chinese during the Han Dynasty was steamed millet served with stew (keng). The keng included various vegetables and could occasionally contain meat. Certainly meat was a regular ingredient for the elite classes. Keng for common people was often only vegetables, but on special occasions chicken meat might be added to the stew. Millet was steamed over the container of simmering keng. Steamed millet would be eaten along with the keng as the basic food in the early part of the Han period.

There is evidence for at least the past 4000 years that millet dough was pulled and stretched to form noodles. During the Han Dynasty a new noodle made from wheat dough, called mien, became a prominent food. Stone mills – powered by animals or water – were introduced in the later part of the Han Dynasty, which allowed large quantities of fine wheat flour to be produced. A mixture of flour and water was also used to make small steamed buns or baked cakes. However, the most important food was noodles. Flattened dough was cut into strips; these noodles could be boiled for immediate consumption. For eating at a later time, especially during a military campaign, the noodles were dried and then boiled only when they were to be eaten.

Chinese families were assigned to work individual farms during the Han period. The labor requirements for these small farms were great. Preparation of the fields, sowing, weeding, distribution of manure, and harvesting on these small farms was almost all done by human power. Consequently, there was considerable pressure on families to produce large numbers of children to labor on the farm and to support the elders in their old age. With improved farming skills during the Han Dynasty, years of good weather resulted in high crop yields and surplus food. The population of China grew at a fairly rapid rate through much of the Han period.

The success in agricultural productivity by the Chinese farmers may actually have contributed to the downfall of the Golden Age of the Han Dynasty. Expanding population and increased food availability freed people to migrate to the cities to work in the expanding imperial government. Consequently, the demands for surplus food production for those remaining on the farms grew at the same time the government increased farm taxes to support the increasing numbers of people now employed by the government. Increasing burdens of heavy taxation led to periodic peasant rebellions, and local warlords gained dominance. Finally in 220 CE, the last Han emperor abdicated to one of these warlords and the Han Dynasty was ended.

## Sources

China Heritage Project, The Australian National University (2005) *China's New Commitment to Heritage (China Heritage Newsletter No. 1, March 2005)*. [Online] Available at: http://www.chinaheritagenewsletter.org/ [Accessed 26 January 2010].

Henry RJ, PS Kettlewell (eds) (1996) *Cereal Grain Quality*. Chapman & Hall, London, UK.

Hsu, C-Y (1980) *Han Agriculture, The Formation of Early Chinese Agrarian Economy (206 BC–AD 220)*. University of Washington Press, Seattle.

Lee MP-H (1969) *The Economic History of China: with Special Reference to Agriculture*. AMS Press, NY.

Lowe M (1968) *Everyday Life in Early Imperial China*. Dorset Press, NY.

Lu H, X Yang, M Ye, et al. (2005) Millet noodles in late neolithic China. *Nature* 439:967.

Smitha F (n.d.) *Map of China: Warring States, 245 to 235 BCE.* [Online] Available at: http://www.fsmitha.com/h1/map08ch.htm [Accessed 26 January 2010].

Trigger BG (2003) *Understanding Early Civilizations: A Comparative Study.* Cambridge University Press, Cambridge, UK.

Wang Z (1982) *Han Civilization.* Translated by KC Chang, Yale University Press, New Haven.

Yu Y-S (1977) *Han China.* In: KC Chang (ed) Food in Chinese Culture, Yale University Press, New Haven.

# 10

# Bantu of Africa

# ~400 BCE to 300 CE

*At the beginning of the world, a being that was both animal and man ...*
*taught the people, with the aid of his staff and his claws, how to change*
*the thorny bush into millet fields.*

Bamana Tribe tale

The ancient history of Africa is often given no more than a footnote in
Western history textbooks. Before the advent of social and cultural history
in the second half of the 20th century, historians commonly focused on the
political and military leaders who left monuments and written records of
their accomplishments: history was written by the victors. The great
cultural revolution of the Bantu in Africa unfolded in a different way
because political and military power were not hallmarks of their society.
Nevertheless, their cultural influence extended over more of the surface of
the globe than nearly any other society in all of recorded history. A major
factor in the unusual historical path of the Bantu resulted directly from

their environmental circumstance and the agricultural practices they developed.

Without extensive written records of the details of this society and their agriculture, knowledge of the Bantu is limited; however it is possible to bring together some of the key aspects of their society. One clear feature of Bantu society is that it did not develop in a single river valley – as was the case of most early agricultural societies. The Bantu had migrated by about 1000 BCE out of the tropical forests of western Africa to the Great Lakes region of eastern Africa (Fig. 10.1). Some of the Bantu established sedentary communities around the lakes, and their food base was derived from these huge fresh-water lakes. Other groups settled in the highlands of the Great Lakes region that at this time included a mixture of forests and savannas. These first Bantu occupying the highlands could continue with traditional hunting and gathering supplemented by growing the African yam.

Yams were no doubt grown by practicing swidden agriculture in which the native vegetation in a small area was removed by cutting and burning to provide an open space for growing yams. The key to the success of the swidden system is that clearing the native vegetation leaves a fairly fertile area in which to grow crops. Ashed residue of burned vegetation provides potassium, phosphorus and other nutrients for the crop, and organic matter from the native vegetation left in the soil provides the essential nitrogen on which crop plants could grow. With continual effort to suppress the return of the native plants to the open space, the crop could be grown for several years depending on the original soil fertility. However, it is likely that the nitrogen level in the soil declined within three or four years to levels incapable of supporting good crop growth, and then reduced yields would dictate that a new plot had to be cleared. This is a sustainable agricultural system if the area originally cleared for growing crops can be allowed several decades to return to native vegetation and reestablish the fertility of the soil.

Climate change was an important factor in the history of the Bantu. Starting in about 3400 BCE the monsoons in Africa no longer penetrated as far north as previously, resulting in an expansion of the Sahara Desert to the south. The growing Sahara Desert forced other people in the region

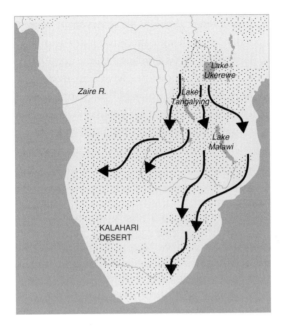

**Figure 10.1.** Map of Bantu expansion in eastern and southern Africa. (Redrawn from Smitha (n.d.).)

south of the desert, called the Sahel, to migrate south to areas of greater rainfall. Eventually, those eastern Sahel people came in contact with the Bantu in the northern Great Lakes region. The people of the Sahel had been growing domesticated sorghum (*Sorghum biocolor* L.) and pearl millet (*Pennisetum glaucum* L.) for some time. In the area that is now Ethiopia, finger millet (*Eleusine coracana* L.) was also being grown. One outcome of the interaction between the Bantu and these peoples from the north was that the Bantu added sorghum, pearl millet, and finger millet to their repertoire of crops. These crops proved to be a very fortuitous acquisition for the Bantu because the climate change causing the expansion of the Sahara Desert also resulted in lower rainfall around the Great Lakes and caused more-savanna-like ecosystems to develop. Sorghum and millets share characteristics that allowed them to be especially attractive for crop production in this new climate. These species have deep roots, giving them access to the reservoir of water held deep in the soil and important for growing a crop in the dry season following the monsoon rains. Also, these crops produce mature grain heads in a fairly short growing season and therefore have an overall lower water requirement per crop. Finally, the seeds of these crops are relatively hard making them well suited for storage over long periods. Sorghum and pearl millet grains held a decided advantage over yams in providing food year round.

---

**Box 9.1.** African Yams

The term African yam refers to a family of plants (*Dioscoreaceae*) that produce edible tubers on their roots. These tubers come in a range of sizes, and some of them have been measured up to 2.5 meters (8 feet) long and 70 kilograms (150 pounds) or more. Yams are rich in starch and have been used as food in Africa for centuries. Yams are usually covered in a thick, rough skin that is difficult to peel. Therefore, food preparation to obtain the starch inside the yam can require considerable work.

African yams appeared to have originated in western Africa where they were called *nyami*. Yams were the staple food in that region. When brought to the Americas as slaves, Africans readily substituted the domestic sweet potatoes (*Ipomoea batatas*) for the yams that were not available in this new land. Sweet potatoes generally have less starch and contain more water than yams. While the sweet potato is an edible tuber with similar general appearance to yams, botanically the plant is quite different from the yams of Africa. Sweet potatoes and yams are members of widely separated plant families. Nevertheless, slaves called the American sweet potato a 'yam', and hence there remains the confusion to this day that the sweet potato eaten in the Americas is sometimes erroneously referred to as a yam.

---

The Bantu may have grown sorghum and millet during two seasons of the year. Both of these species are perennial; that is, another set of plant shoots emerges from the base of the plant after the previous crop is removed. With careful weeding of the crops and timely harvests, a single sowing would produce one to two crops each year of the perennial sorghum or millet over several years. It would not be necessary to sow seeds each year. This, of course, resulted in an important saving in the use of precious seed, making more seed available as food. Once the monsoon season brought rain to the fields, new sorghum and millet plants would emerge from the stumps of the previous crop. Due to the short season required to mature the grains, the first crop could have been harvested before or near the end of the rainy season. This first crop would normally not be water limited, and therefore it would have been the primary crop. In addition, the monsoon could have fully filled the soil with water, allowing a second crop to grow on this stored water into the dry season. Some of the heads of the sorghum and millet could mature on this second set of plants before the stored soil moisture was exhausted.

By approximately 400 BCE, the Bantu had a productive agriculture of mixed crops. The original swidden system of clearing land to grow yams was no doubt extended to growing sorghum and millet. Axes and fire would have been used to clear a new area of vegetation to grow the crops. Seeds would be buried in the fresh, nutrient-rich soil using digging sticks. Men were likely responsible for sowing the crop. After young seedlings emerged, an aggressive effort was required to remove weeds so they would not shade the new crop. Likely all members of the tribe were involved in removing weeds, especially in the first year when native plants would emerge in this new favorable environment. Since sorghum and millet would grow to over 1 meter (3 feet) tall, work to remove weeds shading the crop plants would have diminished as the season progressed. The next challenge would have been to protect the developing crops from foraging animals. This would have been especially important when the grain heads were growing seeds that would be attractive to birds and small mammals. Someone had to guard the plots constantly to scare away any intruding animals. Women and children would have the job of making noise and throwing stones at these intruders. Sentinels would have been especially important during the three or four weeks before crop maturity and grain harvest.

The challenges of removing native vegetation and suppressing invading weeds in the swidden approach to farming were facilitated by the ability of the Bantu to produce iron. It appears based on archaeological evidence that the Bantu developed their own unique system of iron making. The Bantu appear to have been making iron as early as any other society. Iron hoes have been recovered that would have been used for weeding. In addition, a number of iron tips have been found that were likely used as digging sticks to penetrate the soil and sow seeds.

As soon as the sorghum and millet grain heads matured, they would have been harvested from the field and stored. Grains would be removed from storage as needed in the daily preparation of food and beer. The first task in preparing the grain was to grind the very hard sorghum and millet grain. Unlike wheat and barley, sorghum and millet are virtually indigestible without cracking or removing the seed coat. Soaking the grain with wood ash helps to loosen the seed coat so it separates from the rest of the grain. Low acidity resulting from the wood ash not only helped in freeing the seed coat, but it aided in the liberation of several amino acids. Since sorghum grain is low in protein content, the wood ash treatment enhanced nutritional value. The actual cracking of grain was done in large wooden mortars using poles to pound the grain. The pounded grain was sifted through large baskets to separate the larger pieces. These larger pieces of cracked grain could be used directly in making beer, but for making porridge the sorghum and millet grain would be ground further with stone to make fine flour. This flour was added to boiling water in clay pots, and in only a few minutes, the porridge could be eaten. Alternatively, the cooked dough could be cooled and formed into balls. At meals, portions could be torn from the balls and dipped into various sauces.

The prime reason for growing sorghum and millets, however, was often to have grain to produce beer. The ground grain could be placed in water for two or three days to ferment and produce the desired low-alcohol (2 to 3%) beverage. Alternatively, sorghum and millet grains were allowed to sprout in the malting process. The swelling of the grain when it absorbed water would release starch and initiate the breakdown of starch into sugars. When provided with yeast and immersed in water, the sprouted grains could be used directly for fermentation.

The success of the Bantu in growing food using the swidden system resulted in an increase in population beginning in the period around 400 to 300 BCE. This was the beginning of the Golden Age of the Bantu. The increasing demand for food required new lands be used to support the increasing number of people. For a small stable population, it would have been possible to grow crops on one piece of land for three to four years and then cycle to a new part of the surrounding native ecosystem. By extending the cycle over several decades before returning to the same land area, it was possible to sustain production in one area. However, the increasing population of the Bantu required more grain production than possible with the local swidden cycle. The solution was simply to move to new land. The savannas of eastern and southern Africa offered a huge expanse of land that was well suited to the Bantu cropping system based on sorghum and millet. As increasing population required new land for their swidden agriculture, the Bantu moved to new areas to cultivate their staple crops of sorghum and millets on fertile land (Fig. 10.1). The cumulative effect of migration over several centuries to new lands to support sorghum and millet production resulted in the so-called Bantu Expansion. By 300 CE, the peoples of Bantu heritage occupied nearly all land in eastern and southern Africa that was suited for the growth of sorghum and millet.

When the Bantu migrated into a new area, local cultures were suppressed by a mixture of assimilation and by dominance as a result of the superior Bantu technology. The Bantu, however, never became organized into large cities or a unified political force. Possibly this was a consequence of the highly adaptable agriculture they developed based on sorghum and millet, combined with what was a virtually unlimited horizon of new land. The three societies we previously discussed in this book were all seriously constrained geographically. The Sumerians and Egyptians were limited to river valleys surrounded by desert. The Chinese were blocked by dry highland areas to the north and west, and undesirable swamp lands to the south. When population pressures increased in these three societies, their crops and the geography did not offer the option to "move on". These previous societies had to organize for maximum production in the same limited environment. The process of organization resulted in class structures to govern food production and a military to protect/expand the crop-production lands. The Bantu had no such geographical limitation.

The agricultural base of the Bantu allowed them to migrate across Africa but did not result in building of the political or military force to successfully resist domination by outsiders in future generations. Once the Bantu had occupied all the huge land area in eastern and southern Africa in which sorghum and millet could be grown, their grand age of expansion was over. Due to the far-flung area of the Bantu and lack of imminent threat from enemies, no centralized government grew from their cultural expansion. Each of the regions occupied by the Bantu became separated and evolved their own customs and languages. Nevertheless, there is a cultural legacy, particularly in the languages in the areas to which the Bantu migrated. In addition, sorghum and millets are commonly grown in many areas of Africa.

## Sources

Andah BW (1987) *Agricultural beginnings and early farming communities in west and central Africa.* In: BW Andah, AI Okpoko (eds) Foundations of Civilization in Tropical Africa. West Africa Journal of Archaeology, Ibadan, Nigeria.

Ehret C (2002) *An African Classical Age: Eastern and Southern Africa in World History, 1000 B.C. to A.D. 400.* The University Press of Virginia, Charlottesville, VA.

Ehret C (2002) *The Civilizations of Africa: A History to 1800.* The University Press of Virginia, Charlottesville, VA.

Everyday Mysteries, Library of Congress (2009) *What is the difference between sweet potatoes and yams?* [Online] Available at: http://www.loc.gov/rr/scitech/mysteries/sweetpotato.html [Accessed 26 January 2010].

Iliffe J (1995) *Africans: The History of a Continent.* Cambridge University Press, Cambridge, UK.

Schoenbrun DL (1993) We are what we eat: Ancient agriculture between the Great Lakes. *Journal of African History* 34:1-31.

Smitha F (n.d.) *Map of Africa to CE 500.* [Online] Available at: http://www.fsmitha.com/h1/map25af.htm [Accessed 26 January 2010].

University of Pennsylvania, African Studies Center (1995) *Bantu Farming.* [Online] Available at: http://www.africa.upenn.edu/K-12/adib_bantu.html [Accessed 26 January 2010].

# 11

# Maya

# ~150 BCE to 910 CE

*... the creator deities made the first human from white corn [maize] seed that was hidden inside a great eastern mountain under an immovable rock. In order to access this corn seed, a rain deity split open the rock using a bolt of lightning in the form of an axe, and this act burnt some of the corn, creating the other three colors of corn seed: yellow, black and red. The creator deities took some of the freed corn seed, ground it into corn dough, and used it to model the first humans.*

Karen Bassie-Sweet
Corn Deities and the Complementary
Male/Female Principle

The Maya built an impressive civilization beginning about 150 BCE in the area of the Yucatan Peninsula of Mexico and south into Belize and Guatemala (Fig. 11.1). They developed hieroglyph writing to track the history of their civilization, made detailed astronomical observations, and

© T.R. Sinclair and C.J. Sinclair 2010. *Bread, Beer and the Seeds of Change: Agriculture's Imprint on World History* (Thomas R. Sinclair and Carol J. Sinclair)

created a precise calendar to track both seasonal and long-term events. The Maya culture was centered on several large city-states with grand ceremonial complexes and elevated palaces inhabited by the elite of society. Each city-state was ruled from a central city of large plazas and pyramid monuments that dominated the countryside. The central city was surrounded by secondary cities also generally built with plazas and a range of house sizes. Small villages dotted the countryside associated with each city-state. It was the people of these villages who provided food and other services to the leaders.

The topography of the area occupied by the Maya was quite diverse. The region was mountainous in the south and west; in the north were lowlands filled with swamps and shallow lakes (Fig. 11.1). Powerful Maya city-states developed in both areas. The entire region was covered with lush rain forests supplied with rains coming off the Caribbean Seas from June to September. Since the Yucatan Peninsula was underlaid with sinkholes and caverns, much of the water that did not accumulate in the low-lying lakes and swamps would infiltrate to the aquifers that permeated the underlying limestone. As a result, Maya agriculture could not rely on surface rivers for water, but rather the Maya in the lowlands developed a unique cropping system based on the water accumulated in the shallow lakes and swamps.

Due in part to the diversity in the landscape, there were differences among the Mayan city-states in their life styles and food production systems. While there was not a homogeneous Maya culture, common features among the city-states can be identified. A critical commonality among all Maya was their heavy reliance on maize as the basic staple food. One estimate is that at least 50% of the Maya diet was maize. The Maya were the "People of the Corn", as the story at the beginning of this chapter illustrates.

The Maya believed humans were originally made by the gods from maize flour. To associate themselves with maize, the elite inlaid their teeth with jade to emulate the young, green kernels. The heads of female babies were bound to flatten their skulls and take on the appearance of an ear of corn. The leaders of Maya city-states were the personification of the male Maize God and the female Na Goddess, and they exemplified the Maya

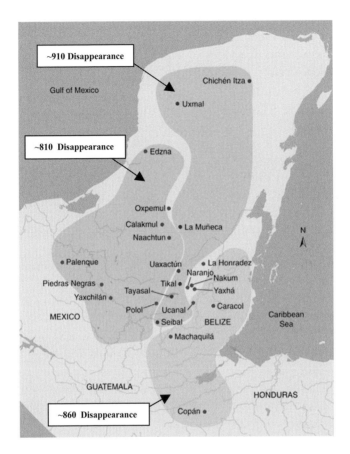

**Figure 11.1.** Map of Maya city-states showing the three regions of successive disappearance of active cities. (Redrawn from Peterson and Haug 2005.)

belief of duality in life. The male and female gods represented the dual structure of the maize plant: the male god symbolized the stalk, tassel, and leaves; the goddess symbolized the ears on the plant. Also, this duality fit with the partnership in marriage in producing maize and the food from it. Husbands sowed, tended, and harvested the crops; wives ground the grain and made the food.

By the time of the Maya, maize had already been domesticated in Mesoamerica for centuries. It had been grown as a crop by the Olmec people, predecessors of the Maya. Maize has many qualities that made it a very attractive crop. Like sorghum and millets in Africa and millets in China, maize is a grass with the extra photosynthetic pathway that lessens its requirements for nitrogen as compared to many other plants. In addition, maize produces a relatively large plant that can out-compete weeds for light if it receives assistance from humans by removal of its largest plant competitors. Also, the number of grains produced on a single plant can be quite large, so the ratio of harvested grain number to sown grain number can be large. The tropical rains in June to September provided the water to readily grow maize. However, during the very dry season of February through May it was likely that cropping was impossible.

The early Maya no doubt practiced swidden cropping, removing the native rain forest vegetation in small patches to grow their crops. Likely, maize was not grown as a monoculture but as a mixture of maize, bean, and squash – the American "trinity". Bean has the advantage of being able to carry out symbiotic nitrogen fixation so nitrogen is obtained directly from the atmosphere. Squash is a low growing crop with large leaves. When squash plants grow under the maize plants, the large leaves of squash shade the soil surface suppressing weeds, and protecting the soil surface from erosion by pounding rains, and reducing high water evaporation rates from the soil surface. After three or four years the cleared patches would be allowed to return to rain forest, and a new area would be cleared. The time interval before returning to the same area would have been several decades when the demands for food production were low. The rapid return of the forest to the cleared area allowed restoration and replenishment of the soil.

The swidden system would have readily provided nutrients the maize crop required. However, climate and topography restricted the geographical spread of the Maya. When increasing population resulted in the need for greater production, the Maya increased cropping land within their geographical confines. One solution was to terrace sloping lands, especially in the highlands, so that crops could be grown on the hillsides.

Terraces were formed by using rocks to support narrow strips of flat land on which maize, bean, and squash could be grown. Terraces helped to trap water from the summer rains and decrease soil erosion. Water was retained on each terrace so that the water could soak into the soil. Terracing to some extent was practiced by many Maya communities.

The Maya also developed a unique farming system that allowed them to expand their crop productivity into wetland areas of swamps and shallow lakes that formed in the rainy season. The Maya created raised beds, called chimapas, in these wetland areas on which to grow their crops. The native vegetation was removed from the wetlands and piled into beds. Organic matter collected from the bottom of the swamps and lake beds was added to the chimapas. These beds were commonly two to ten meters (six to thirty feet) wide and separated by water channels that allowed the water to flow throughout the array of beds. New material had to be continually added to the beds to keep their surfaces above the highest water level so the crops would not be flooded.

The chimapas were well suited for growing maize. Water was readily available to the plants during the wet season as it infiltrated into the chimapas. In fact, one worry may have been to construct the chimapas high enough so that a major tropical storm would not swamp the bed and flood the plants. In fact, the Maya could have selected flood-tolerant lines of maize which might have had roots containing aerenchyma, small air channels in their roots that allow air to reach the root tips and keep them alive. If the chimapas were not flooded for too long a period, the maize plants would be unharmed.

In addition to a ready supply of water, a key feature of the chimapas would have been their ability to provide the growing crops with nitrogen. New organic material would be added to the bed by transferring settled plant materials from the bottom of the adjacent water channels to the top of the beds. Much of this new material would be organic matter washed into the swamps and lakes from the adjacent rain forest. New organic materials added to the beds each year would decompose when exposed to air. Oxidation occurs rapidly in warm, wet environments so a steady supply of nitrogen for the maize plants would have been available from the organic matter added to the chimapas. The water and nitrogen environment of chimapas would have been nearly ideal for maize. However, oxidation of the organic matter resulted in an annual subsidence. This subsidence would have required a highly structured society to provide the labor for adding new material to the chimapas each year.

Weeding would likely have also been a major activity in growing maize in the chimapas system. Ready access to water and nitrogen on the chimpas beds would have encouraged the growth of undesired weed plants to compete with the maize. Further, each new addition of organic matter to the chimapas beds added a new supply of weed seeds. Control of these new weed seeds would have been facilitated if the additions to the chimapas beds were done early in the dry season. Weed seeds would readily germinate in the wet conditions on top of the chimapas, but the young seedlings would emerge into dry conditions. With a receding water table around the chimapas, many of the weeds would likely be killed by the end of the dry season, before the maize was sown.

Weeding at the beginning of the cropping seasons with the return of the rains would have been an especially arduous task. There were no animals or tools to assist with the weeding, so hand labor was needed to remove the weeds. During the time of hot and extremely humid weather, there would have been an abundance of flying insects, including mosquitoes. Fortunately, the tall-growing maize would have required removal of only the taller-growing weeds that would shade the maize plants. Squash plants below the maize would also help to shade out emerging weeds.

Due to the extreme of the wet and dry weather cycle in which the Maya lived, their activities had to be well tuned to the seasonally changing conditions. It is not surprising that the Maya developed a precise calendar to track these cycles, especially when two successive crops were to be squeezed into the limited time when water was available. The Maya believed the crop had to be sown at the full moon closest to the beginning of the wet season in June. The maize seeds were separated into four colors and sown in the four quadrants of the field symbolizing the daily cycle of the sun: red in the east, yellow in the south, black [actually purple] in the west, and white in the north. A digging stick was used to create a hole into which seeds of corn were sown; usually five seeds would be placed in the hole. Holes in the four directions meant sowing seeds in units of 20, the base number in Maya counting.

The crop sown in June would likely have matured in October. Once the ears were near maturity, the Maya would bend the maize stalk over just below the ear. This caused the ear to point downwards and helped minimize losses during the ongoing rainy season. The resulting shorter height of the plants would make them less vulnerable to winds and the downward orientation of the ears helped keep rain water from seeping into the husks. These plants could stay in the field until the ears could be collected and taken to the village for storage.

There is no archaeological record of the full cropping cycle used in the production of maize on the chimapas. However, it is possible that two crops a year might have been grown on chimapas beds. Even after the rainy season ended, swamps and lakes would have been filled with water. Depending on the rainfall in the transition months between the rainy and dry seasons and the drainage rate of these water bodies, a high water table may have persisted sufficiently long that maize roots could access enough water to at least produce immature but edible "green" ears.

The chimapas needed to be reconstructed soon after harvest. However, there would have been little else to do in tending the fields during the four dry-season months, i.e. roughly February through May. Like the cropping cycle in Egypt, this break in the cropping season would have released a large labor force available to build the massive monuments constructed by the Maya.

Mature ears of maize brought from the swidden fields, terraces, and chimapas were stored until required in food and beverage production. When grain was to be consumed, the kernels would be removed from the cob and soaked in water. The maize of the Maya was a "flint" seed with a rounded shape and very hard seed coat. Flint seed contrasts with most modern maize, which is a "dent" seed identified by the indentation at the end of the seed allowed by the softer seed coat. The hard seed coat of flint corn was an advantage for the Maya in protecting the seed from being attacked by various pests once the grain had matured.

A disadvantage of the flint seed was that the hard seed coat had to be removed from the grain before it could be consumed. The Maya soaked the grain in water to loosen the seed coat. To aid in loosening the seed coat, the Maya made the ingenious discovery that adding lime or ash to the water expedited the process. The lime was obtained from natural limestone formations or from seashells. Likely unknown to the Maya, the lime had a major nutritional benefit. Lime neutralized the phytate in the corn that blocks iron and zinc absorption in the human intestine. Also, the lime soaking liberated several amino acids, including tryptophan, in the maize grain that would not otherwise be available for human absorption. Most importantly, the lime treatment made niacin available in digestion; the absence of niacin can result in pellagra. While pellagra has been a major disease in the last two centuries in some societies that relied heavily on maize consumption, it was not a disease of the Maya. Additionally, soaking the grain in lime meant that calcium intake was high, and the Maya experienced little osteoporosis.

Once the seed coat was loosened, water was drained from the grain. The wet grain was then ground by hand between two stones. The resulting paste was mixed with water to make a thick gruel called pozole. The paste was alternately formed into flat cakes and roasted on a pottery griddle to make tortillas. Soaked beans were mashed and wrapped in tortillas to form a food approximating a modern burrito. The addition of beans substantially increased the protein of this food, including niacin, tryptophan, and lysine. Chilies were likely added to the bean mix to spice the flavor.

The Maya also ate tamales, which were made from a mixture with the maize dough that might include beans, chili sauce, and meat from hunted animals. The mixture was wrapped in green maize leaves, or banana leaves, and steamed. Combining maize with meat and vegetables grown in garden plots and collected from the wild gave the Maya a well-balanced diet. Squash and the seeds of squash were also a major component of their diet.

The carbohydrate level of maize allowed it to be readily fermented, and the Maya consumed several fermented products. Pozol, maize dough, can be formed into small balls and wrapped in leaves. The wrapped pozol will ferment in anywhere from one to fourteen days to yield a fermented food containing alcohol. This fermented maize ball can have some useful medicinal properties. Since fungi also grow on the maize ball while it is fermenting, antibiotics are produced, and the balls could be used to treat infections and wounds.

Beer – chicha – was certainly an important part of the Maya diet. Chicha was made from dry maize kernels ground between stones. The maize flour was mixed with hot water to aid in the breakdown of the starch. The mixture was kept hot for about one hour and then allowed to cool. With cooling the particles settled, and a liquid top layer formed. The top layer of the cooled mixture was removed and placed in pots for fermentation. In two to six days the fermentation was complete, and the chicha was ready for consumption. Since the fermentation pots were not cleaned, the yeast was passed from one batch of chicha to the next.

A variant of the above procedure to enhance the sugar yield of the maize was to add an enzyme to the maize flour before fermentation. In this case, the enzyme was salivary amylase, which is present in abundance in human saliva (Box 3.1). Maize flour was slightly moistened and formed into a small ball. The ball of flour was popped into the mouth and thoroughly mixed with saliva using the tongue. The mixture was pressed against the roof of the mouth to form a small disc. The salivary amylase would break down the starch as the discs were dried in the sun. When the discs were to be fermented, they were ground and used as the raw material for fermentation as described above.

Maize provided the basic staple food, which when combined with beans and squash provided a solid diet for the Maya. This diet allowed the population of the Maya to grow to reach something on the order of two to three million people. As described above, the need for increased food supply was met by expanded cropping on terraces on sloping lands and the chimapas in the lowlands. However, these schemes were particularly vulnerable to low rainfall. The key to highly productive terrace and chimapas systems was the reliability of the summer rains. For the terraces, rains were necessary to fill the soil with water at the end of the dry season so a new crop could be grown on the sloping land. Unless rains refilled the soil, there was little source of water in the highlands on which to grow the crop. For the chimapas, rains were required to fill the swamps and lakes with water and carry organic matter into the channels between the raised beds. Without rains the first maize crop would suffer from a poor water supply, and a second crop would not be possible. Also, decreased rainfall limited the growth of the surrounding native vegetation, which was essential in providing the organic matter added to the chimapas for maize production in following years. Low water levels in the channels between the chimapas beds over extended periods would allow rapid oxidation of the organic matter in the beds and cause a loss of nitrogen. The drying of the wetlands meant a dramatic loss of productivity of the maize crop due to inadequate water and nitrogen.

Episodes of anomalous weather occurred in Mesoamerica beginning in about 750 CE with specific drought events over the following 150 years. Droughts are inferred from sediments collected in core samples taken offshore of Venezuela. These cores indicate that there were three major drought periods in Mesoamerica coinciding exactly with the disappearance of the Mayan societies. The first drought was at approximately 810 and lasted eight years, coinciding with the disappearance of Maya city-states in the west furthest from the Caribbean coast. The second drought was at about 860 and lasted about three years, coinciding with the loss of city-states in the southeast. The last drought was about 910 and lasted six years, coinciding with the disappearance of the Maya city-states in the central and northeast regions. These events certainly seem to support a suggestion that climate perturbation played a major role in the collapse of the Mayan society.

Of course all the factors causing the disintegration of the cities will probably never be known, but it is hard to imagine that droughts were not major contributors. When the royal personification of the Maize God and the Na Goddess failed to deliver the rains and abundant maize yields, the fundamental core of the whole Maya social structure would have been shaken. The increasing populations in the central cities would have been especially vulnerable to low food availability and surely applied pressure to the central authority that may have led to its collapse. Those living in the villages could have managed subsistence production and relied once again mainly on swidden production for survival. Those who successfully survived in the rain forest could have become the ancestors of today's people of Maya heritage. However, the formerly great Maya city-states literally disappeared under the re-growth of forests.

## Sources

Bassie-Sweet K (1999) *Corn deities and the complementary male/female principle.* Presented at La Tercera Mesa Redonda de Palenque. www.mesoweb.com/features/bassie/corn/media/Corn_Deities.pdf [Accessed 26 January 2010].

Benson EP (1977) *The Maya World.* Thomas Y. Crowell, NY.

Hillel DJ (1991) *Out of the Earth. Civilization and the Life of the Soil.* The Free Press, NY.

Kiple KF, CO Kriemhild (2000) *The Cambridge World History of Food, Volume One.* Cambridge University Press, Cambridge, UK.

Lorence-Quinones A, C Wacher-Rodarte, R Quintero-Ramirez (1999) *Cereal fermentation in Latin American countries. In: Fermented Cereals.* A Global Perspective, Food and Agriculture Organization of the United Nations Agriculture Services Bulletin No. 138, Rome.

Peterson LC, GH Haug (2005) Climate and the collapse of Maya civilization. *American Scientist* 93:322-329.

Turner BL II (1990) *The rise and fall of population and agriculture in the central Maya lowlands: 300BC to present.* In: LF Newman (ed) Hunger in History: Food Shortage, Poverty, and Deprivation. Basil Blackwell, Inc., Cambridge, MA.

White CD (1999) *Reconstructing the Ancient Diet.* The University Press, Salt Lake City, UT.

# 12

# Athenians/Romans

## 550 to 338 BCE/509 BCE to 410 CE

*What now remains of the formerly rich land is like the skeleton of a sick man, with all the fat and soft earth having wasted away and only the bare framework remaining. Formerly, many of the mountains were arable. The plains that were full of rich soil are now marshes. Hills that were once covered with forests and produced abundant pasture now produce only food for bees. Once the land was enriched by yearly rains, which were not lost, as they are now, by flowing from the bare land into the sea. The soil was deep, it absorbed and kept the water in the loamy soil, and the water that soaked into the hills fed springs and running streams everywhere. Now the abandoned shrines at spots where formerly there were springs attest that our description of the land is true.*

Plato

The Athenians and Romans made huge contributions to Western culture, and each, of course, is deserving of in-depth historical analyses. In the context of the intertwining of history and agriculture and food, these two societies had many similarities. The nature of the environment in which they developed and the subsequent historical progression of their food supplies followed similar patterns. Grain supply evolved from abundant local production, to destruction of the local environment for grain production, and finally to nearly total dependence on imported grain supplies. While the Romans were able to prolong the whole cycle for a much longer period of time than the Athenians, the basic ecologically destructive pattern was the same for both societies.

In the early years before their Golden Ages, the soils of both Attica – the farming region around Athens – and the farms near Rome were quite productive. People farmed relatively small farms to feed their families wheat and barley, which originated in the Fertile Crescent. These small farms produced sufficiently large quantities of grain to free some people from growing crops and allow activities not directly related to food production. Athens and Rome developed from the small non-farm population that collected in cities and engaged in politics, commerce, and the arts.

As the population of Athens and Rome grew, the countryside became heavily exploited to provide food for the cities. The response in the countryside to the food demands of these two cities was similar: maximize production without regard to the environment. Athenians and Romans shared a common philosophical view of nature. Both considered nature a resource to be controlled and used for man's benefit. The superior intellect and stature of humans put them in the rightful position of dominion over nature. This view led, of course, to a very exploitative approach to their environment. The hillsides were cleared of trees for wood to construct buildings, make furniture, and provide the main fuel in cooking and heating. By the beginning of the Golden Age of each society nearly all the local usable timber had been cut. Any remaining shrubs and herbs were used to graze animals. Those lands sufficiently flat for cropping were burned to remove the last of the native vegetation and then plowed for sowing crops.

Unfortunately, the topography and soils of the lands surrounding Athens and Rome were especially vulnerable to damage from intensive cropping. The land is hilly to mountainous without extensive flat areas for crop production. The native vegetation in these regions was a mixture of trees and shrubs that grew on shallow soils. The soil was loam or clay-loam commonly less than one meter (three feet) deep that had developed on top of limestone bedrock. In its native condition, the loamy soil had the capacity to absorb the winter rains and provide a moisture reservoir sufficient for native plants into the dry season. Once eroded, however, there was no longer soil with capacity to retain water.

Athens and Rome are in the Mediterranean climate zone, which has two distinct seasons: wet winters with intense rains followed by very dry summers. (This is the same climate that exists in the Taurus Mountains where the winter rains fed the Euphrates River to give high early spring flow in Sumeria.) The winter rains often come in heavy squalls with intense rainfalls that would readily carry away the exposed top soil. Consequently, winter rains resulted in considerable amounts of water running off the land carrying with it soil stripped from the hills and mountains. The loss of this soil eventually exposed the underlying limestone, which resulted in the rocky outcrops of limestone that are the characteristic landscapes of modern Greece and Italy. The eroded soil was swept into valleys where the soil and water accumulated to form marshes that could not be farmed.

The erosion of hillsides and filling of valleys with soil caused lands to be abandoned for production of cereal grains. The Greeks and Romans were well aware that changes were happening in the landscape and that fields were becoming less productive. The statement by Plato that introduces this chapter succinctly describes the situation he found in Attica. However, the Athenians and Romans simply concluded their land was "growing old", and they needed to move to new lands to obtain their grains. The greatly diminished capacity to produce grain at home required both the Athenians and Romans to acquire colonies, or at least control foreign lands to provide the grain needed by the homeland. The course of history for these two societies was to a large extent dictated by the loss of self-sufficiency in food production capacity which dictated military action to obtain and control grain production in distant lands (Fig. 12.1).

As soil erosion took its toll on the crop land, there had to be a shift in cropping practices. As in Sumeria, barley became a key "backstop" crop, particularly in Attica. In these regions, the key advantage of barley is that it has a shorter growing season than wheat; it reaches maturity earlier in the season. This was important as the capacity for soil-water storage was diminished by soil erosion. When there was less water to carry the crop into the dry season, a crop that matured before severe dry conditions developed in the soil would have an advantage. By a wide margin barley became the major grain crop produced in Attica. Barley was consumed as bread and beer by slaves and peasants and by the Athenian armies when on military campaigns.

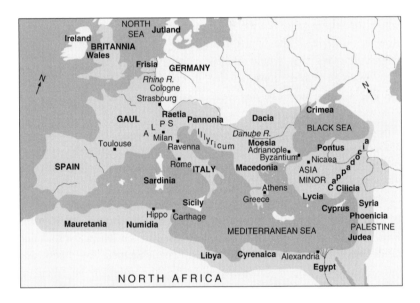

**Figure 12.1.** Map of the Mediterranean Sea region. Shaded area shows the extent of the Roman Empire at its peak. (Redrawn from Smitha (n.d.).)

Barley, however, was never accepted as a major crop in Rome. Barley is cited in Roman writings as being fed to those soldiers who had performed poorly in battle, and the punishment of eating barley could last for years. Also, the Romans did not use barley in making large quantities of beer since they much preferred fermented grapes. On military campaigns, however, it was not possible to carry large amounts of wine, so grains were transported both to bake bread and also to ferment as a beverage. Wheat was always the main staple grain of the Romans. When wheat was no longer produced in quantity in areas surrounding Rome, consumer demand continued and even increased, so the city became nearly completely dependent on grain imported from her colonies.

Wheat consumed as the staple crop of Athens and Rome was usually not the emmer wheat produced by earlier civilizations. Bread wheat (*Triticum vulgare*) had been selected as the favored wheat by the time of the Athenians and Romans. Bread wheat was much easier to thresh than emmer wheat because the grain hulls were not tightly attached to the seed. In addition, bread wheat was lower in protein content and more glutinous than emmer wheat so it could readily rise to make a light, leavened bread.

The early farmers in both Athenian and Roman societies focused on the production of bread wheat in rather small plots. Wheat may have originally been grown in a two-year rotation: one year of production and one year of fallow allowing soil rejuvenation. The fallow period would have given the soil an opportunity to mineralize more nitrogen so that it would be available for the next wheat crop. As the populations grew and the demand for grain increased, however, the fallow system was abandoned in favor of continuous grain production. Weed control between successive crops would have required intense plowing. The Roman regime was to plow the field early in the spring, again during the summer, and a final time before sowing the crop. The loss of the two-year rotation and continual working of the soil sped the loss of soil by water erosion.

Oxen were used to pull iron-tipped plows through the fields of Attica and Rome. The iron tips extended the lifetime of the plows in working the rocky soils. The use of oxen in these societies marked the first major replacement of human muscle by animal muscle in crop production. This was possible because animals could be grazed on the hilly lands not used in growing of crops. Consequently, it was not necessary to grow crops to specifically feed animals, except for the collection of some hay for feed during the driest part of the summer. The grazing animals would usually be gathered each night into stables and provided bedding of straw previously collected from the grain fields. The manure of the animals would be retained by the straw and eventually removed from the stable and applied to crops. The Romans, in particular, were very much aware of the benefit of returning animal manure to the fields to enhance the performance of subsequent crops.

---

**Box 12.1.** Wine in Athens

The combination of loss of local grain productivity and increase of wealth in the city encouraged increased wine production and consumption by the Athenians. Improved grapes (*Vitis vinifera*) for wine production were derived from native varieties and from vines brought from Asia Minor. Grapes are a hardy plant that could be planted on eroded hillsides. Grapes, and additionally olives, became the dominant crops in Attica since they were the most profitable for producing wine, olives, and oil for sale in Athens. Olive trees are also well adapted to survive in climates with limited water availability. The Greek historian Thucydides wrote that "the peoples of the Mediterranean began to emerge from barbarism when they learned to cultivate the olive and vine".

The white grapes of Attica were harvested in the fall to make wine. Grapes were brought to large wooden or earthenware vats where workers would crush the grapes with their feet. The juice was then placed in large jars for fermentation. The juice was high in sugar content, giving an alcohol content of the wine as high as 15 or 16%. Grapes were also boiled to release more sugars and add to the sweetness of the wines. Greeks enhanced the flavor of their wines by adding resin, herbs, spice, seawater, brine, oil, and perfume. The wines were stored in large amphorae. Oxidation was a common problem in storage, so most wines were consumed within a year of harvest.

The "civilized" method for consuming the high alcohol-content wine was to dilute it with water. The ratio was commonly two to four parts water to one part wine. Wine was felt to be medicinal and was thought to cure a number of ailments. However, much wine was consumed in social events called "symposia" at which men gathered to discuss a wide range of topics. In the Athenian philosophy, symposia were democratic as all participants (only male, of course) shared the same wine and were given an opportunity to talk. The challenge for a participant was to manage the wine dilution so that his ideas were freed in the brain but his tongue remained lucid.

**Box 12.2.** Wine in Rome

White wine was the essential beverage for consumption by Romans. By about 100 BCE, beer consumption had virtually disappeared and wine was consumed by all levels of society. At the peak of Roman economic power, nearly 190 million liters (47 million gallons) of wine were consumed each year. This represents about half a liter per day for every man, woman, and child in Rome. Grapes were grown in the countryside near Rome and much of the wine was fermented locally, but still large quantities of wine needed to be imported. Much of the best wine was obtained from grapes grown in the areas south of Rome. The disastrous eruption of Mt. Vesuvius was especially catastrophic because it destroyed some of the finest vineyards and wineries.

The Romans were efficient with their grapes in extracting every bit of material from the grapes for wine making. Top quality wine was that made from the first juice extracted by workers stomping on the grapes. This juice had the highest sugar content, and these wines were the most costly and consumed only by the elite classes. The next lower quality of wine was derived from the juice squeezed from crushed grapes by a mechanical press. Grapes were placed between wooden beams and juice extracted by squeezing the grapes. The initial juice from the mechanical extraction was, of course, less sweet and was sold to the second tier of Roman citizens. Honey or other sweeteners were often added in quantities to this wine to enhance its flavor. Further squeezing by the mechanical press produced a sour wine that was low in alcohol content. This third tier of wines was provided to the soldiers due to its low alcohol. The lowest tier of wines was obtained by soaking grape skins and seeds left from the squeezing. This sour, weak wine was given to slaves and the poorest people. Over time, oxidation of the alcohol in the weak wines would result in vinegar, which is likely the drink the Roman soldiers gave to Jesus during his crucifixion.

As in other activities, the Romans borrowed heavily from the Greeks in their viticulture, wine making, and consumption. The "civilized" practice of diluting wine was continued by the Romans, although the Romans often decreased the dilution to only two parts water to one part wine. The aqueducts brought fresh water to allow safe dilution of the wine. Much wine was consumed in "convivia" that originated from the Athenian idea of a symposium. However, the

**Box 12.2.** Wine in Rome *continued*

Roman convivium evolved away from the Athenian ideal of democratic behavior to an event for displaying social status. People of higher status held convivia inviting people of equal status and also those who had obligations to the host and upper-ranked guests under the patronage system. Invitees were seated in the convivium by rank, making the status of individuals explicitly clear. The food and wine served to guests also varied by each individual's status. Rather than engaging in extensive discussions of philosophy, Romans opted for entertainment that included insulting toasts to those of lower status. The convivia eventually digressed into orgies in the later stages of the degeneration of the Roman Empire.

The classical Roman agricultural writers Pliny and Commella suggested more extensive rotations than the simple two-year rotation. They recommended the addition of a third year of a legume crop to restore the soil, which we now know would result in a replenishment of nitrogen to the soil. There is, however, little evidence that such a three-year rotation actually gained widespread practice. Under the pressures to generate profits for city-dwelling owners, fields were not left idle for a fallow year or planted in a crop that did not produce immediate income.

Grain was harvested on the farms of Attica and Rome using age-old procedures. Grain heads were removed from the plant stems using iron sickles. Threshing the grain was done on a threshing floor usually constructed on an elevated site exposed to steady wind. The threshing floor could be constructed of stones or hard-packed clay. Slaves using sticks or flails would beat the grain heads to remove the grain. An alternative was to have animals, particularly oxen, walk on the heads to release the grain. In some cases, the oxen pulled a sledge over the heads to further aid in separating grain from the heads. The chaff was removed from the grain/chaff mix by tossing the threshed material up into the wind so the chaff would be blown away. Of course, in the later days of the

Roman Empire this threshing was done in distant lands and only the grain was shipped to Rome.

As erosion progressed on the lands around Rome and wheat production became difficult, there were major shifts in agriculture. Lands became consolidated into large holdings called latifundia. The latifundia were owned by rich patricians who commonly lived in Rome. Overseers ran the estates, often with brutality to force peasants and slaves to do the arduous work of growing and processing crops. The domestic products that made the most money in the markets of Rome were wine, olive oil, and meat. Hence, the amount of land devoted to wheat was greatly decreased, and some of the latifundia themselves may not even have been self-sufficient in grain production. Olives and grapes were less demanding of water and nutrients than wheat. The transition to these products being demanded in Rome suited the eroded and low-nutrient soil. Manure applications could be concentrated at the base of olive trees and grape vines rather than being broadcast across an entire wheat field. Further, the meat produced on the latifundia was based on animals grazing on hillsides unsuitable for growing crops. The widespread de-emphasis by Romans of staple-grain production in favor of "commercial" crops was a major change in perspective concerning their dependence on imported grains.

Porridge made from the ground flour of barley in Athens and wheat in Rome was the most common way of consuming grain in the early days of these cultures. The grain was ground in their homes using traditional techniques, i.e. using a saddle quern or mortar and pestle. However, by about 200 BCE the Romans had developed a rotary hand quern for use in the home to grind the grain. The rotary quern was made of two stones between which the grain was ground (Fig. 12.2). The top stone rested on a wooden pivot post mounted in the center of the bottom stone. Grain for grinding was introduced through a hole in the top stone. A handle mounted on the top stone allowed it to be rotated to grind the grain. The ground flour trickled out the edge of the quern.

As the grain demand grew in Rome and large quantities of flour were required, the grinding of grain moved out of the home and became a specialized trade. The hand rotary quern was replaced by a much larger version of similar design. Grooves were cut into the millstones to achieve a finely ground flour for making bread. The large top stone in this grain mill was turned by pushing a large horizontal piece of wood attached to the pivot above the top stone. Slaves or animals provided the power for grinding the grain by continuously walking around the mill and pushing the mill yoke. These larger mills allowed substantial amounts of grain to be ground each day in a single Roman mill.

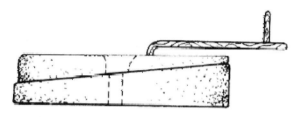

**Figure 12.2.** Hand rotary quern used by Romans to grind grain. (From Penn 1966, courtesy of W.S. Penn.)

A further modification to the Roman mill was to replace the power provided by slaves or animals with water power. A paddle wheel was attached to the grinding stone by a series of gears, then water from a stream falling on the paddle wheel would cause the top stone to move. However, there is little evidence of many water mills having been constructed in Rome; it appears the Romans generally found it more expedient to use slave labor in the grain mills.

In later times in the Roman Empire, bread replaced porridge as the most important part of the Roman diet. Bread was the food of "citizens" of Rome, a status commonly achieved by successful military service. In later days of the Roman Empire, bread was distributed as a right to the citizens of Rome. The baking of bread was done, not in the home but rather in specialized bakeries. The danger of fire spreading when baking bread in a home oven precluded baking in each home or tenement in the tightly packed city. Roman politicians used bread from the centralized bakeries as political capital to buy votes of the plebeians.

The growing demand for food in both Athens and Rome along with the decreasing capability of cereal production on the eroded homelands meant that neither society was able to remain self-sufficient in food production. Both societies had to turn to importation of huge quantities of grain from outside of the local region. The Athenians turned to nearby islands in the Aegean Sea as major sources of wheat. As the demand for grain increased, grain had to be obtained from farther distances. Athens became an unrivaled sea power controlling the eastern Mediterranean Sea to protect the grain supplies from pirates and foreign enemies. Ultimately, much of the wheat used by the Greeks was shipped from Egypt and the region of modern Ukraine and Romania via the Black Sea. Control of the Bosporus by the Athenians ensured their control of the flow of grain in the eastern Mediterranean.

The ability of Athens to control colonies ended in 338 BCE with its military defeat by the Macedonians. Grain continued to be imported from the Black Sea regions, but the movement of these shipments was no longer under the control of Athens. The power and influence of Athens had been lost. The Macedonians, and in particular Alexander the Great, took command of a new Empire.

The Romans, similarly, became fully dependent on wheat grown in far-flung colonies. It was necessary to conquer new lands that grew the bread wheat needed in Rome. The Romans first fought the Punic Wars with Carthage gaining the wheat output of Sicily and north Africa. They next defeated the Macedonians in 168 BCE at Pydna gaining control of the eastern Mediterranean and the grain coming from the regions bordering the Black Sea. Julius Caesar added the rich grain output of Egypt, Gaul, Spain, and Britain to the list of colonies supplying grain to Rome. At the height of the Roman Empire, its borders were expanded to encompass all accessible lands suitable for production of wheat. The far-flung array of colonies now provided to Rome the flow of wheat that it could no longer grow itself.

While the Roman Empire survived for centuries, its expansion to widely dispersed wheat-production lands brought its borders to areas occupied by non-agricultural tribes and necessitated the garrisoning of soldiers throughout the Empire to ensure the protection of the grain supply. There were continual conflicts with these tribes, particularly Germanic tribes in the north, eroding the power of Rome. Also, there was the challenge to sustain the far-flung supply lines of wheat to feed the population of Rome. Eventually, Rome was unable to maintain control of its "wheat empire", and the Empire collapsed.

## Sources

Braun T (1995) *Barley cakes and emmer bread.* In: J Wilkins, D Harvey, M Dobson (eds). Food in Antiquity. University of Exeter Press, Exeter, UK.

Curtis RI (2001) *Ancient Food Technology.* Brill, Leiden.

Dale T, VG Carter (1955) *Topsoil and Civilization.* University of Oklahoma Press, Norman, OK.

Fussell GE (1966) *Farming Technique from Prehistoric to Modern Times.* Pergamon Press, NY.

Garnsey P (1988) *Famine and Food Supply in the Graeco-Roman World.* Cambridge University Press, Cambridge, UK.

Harlan JR (1992) *Crops & Man.* American Society of Agronomy, Madison, WI.

Hillel DJ (1991) *Out of the Earth. Civilization and the Life of the Soil.* The Free Press, NY.

Montanari M (1994) *The Culture of Food.* Translated by C. Ipsen. Wiley-Blackwell, Oxford, UK.

Penn WS (1966) *A Quern Survey in Kent.* Kent Archaeological Review, Spring 1966 (Issue #3). [Online] Available at: http://cka.moon-demon.co.uk/KAR3AND4/KAR3AND4_quern.htm [Accessed 26 January 2010].

Phillips R (2000) *A Short History of Wine.* Harper Collins, NY.

Ritchie CIA (1981) *Food in Civilization. How History has been Affected by Human Tastes.* Beaufort Books, Inc., NY.

Robinson J (ed) (2006) *The Oxford Companion to Wine, Third Edition.* Oxford University Press, Oxford, UK.

Smitha F (n.d.) *Map: Roman Empire, CE 150.* [Online] Available at: http://www.fsmitha.com/h1/map19rm.htm [Accessed 26 January 2010].

Solbrig OT, DJ Solbrig (1994) *So Shall You Reap: Farming and Crops in Human Affairs.* Island Press, Washington, DC.

Sonnenfeld A (1999) *Food: A Culinary History from Antiquity to the Present.* Translated by C Botsford. Columbia University Press, NY.

Tannahill R (1973) *Food in History.* Penguin Group, UK.

# 13

# Feudal Europeans

# 800 to 1347

*A householder and that a great was he;*
*Saint Julian\* he was in his country.*
*His bread, his ale, was always after one.*
*A better envined man was never none.*

[\*patron saint of hospitality]

Geoffrey Chaucer
Franklin's Tale in
*Canterbury Tales*

Following the dissolution of the Roman Empire, Europe fell into disarray and the countryside was again divided among small fiefdoms. The church in Rome enjoyed an increase of power and wanted to consolidate political authority. Centralized authority was vested in Charlemagne in 800 when Pope Leo III crowned him emperor of the Holy Roman Empire. The Holy Roman Empire encompassed all of central Europe (Fig. 13.1) and recognized the central authority of the Pope. In reality, however, military power remained in the hands of the lords of feudal kingdoms. A further splintering of power occurred within these kingdoms as local nobles controlled areas protected by their castles, the remnants of which still can be seen dotting the European landscape. The feudal structure of the Holy Roman Empire was maintained for more than half a millennium – even through the Black Plague that decimated Europe beginning in 1347.

**Figure 13.1.** Map of the extent of the Holy Roman Empire. (Redrawn from Smitha (n.d.).)

The feudal period of the Holy Roman Empire is sometimes referred to as the High Middle Ages, and for agriculture and foods this was a period of considerable technological progress, even though the pace of development was slow. New cropping techniques and food preparation methods were introduced and developed during this period. The forests of central Europe, which had intimidated the Romans, were slowly removed and replaced by crop lands. The increase in crop land, expanded production of new crops, and changes in foods provided the base for a population expansion during this period. Specialized trades such as millers, bakers, and brewers – all associated with handling the grains – evolved during this time. Also, a small fraction of the population was being freed from the labors of growing crops and producing food to work in the visual and literary arts. The seeds of the Renaissance were being sowed at this time; however, their growth was to be dramatically interrupted by the Black Plague.

Castles and their surrounding villages were the centers of the agricultural system, which was under the absolute control of the local nobles and lords. All activities, including decisions on growing crops, ultimately emanated from the nobles. Peasants working the fields were bound so tightly to the castle for protection and use of land that they were in effect slaves to the feudal system. Peasants had no choice but to work the land in exchange for a place to live, a small share of the grain produced from their labors, and garden land to grow vegetables. In exchange the peasants received protection in the castle when invaders threatened the countryside. Also, the system was designed to keep the peasants fully indebted when grain had to be "borrowed" from the noble in years of poor yields. Since debt was passed from father to son, not even death allowed the family to escape the control of the noble. A peasant born owing a substantial debt to the noble had no hope of escape.

Much of the land in central Europe at the beginning of this era was still in forests and grasslands. Unlike climates in most of the societies previously discussed, Europe's climate did not include a major dry season; therefore, lack of water was not a serious concern in most years. A long cold winter and a fairly short summer were typical across much of the Holy Roman Empire. Specialized crops had to be exploited to match this climate. Bread wheat varieties developed from those used by the Romans were grown as a winter crop in the southern portions of Europe. In the most northern parts of Europe the winter was sometimes too brutal to allow even winter wheat to grow, and varieties were selected to be sown in the spring and grown through the summer. Summer wheat, however, was vulnerable to summer frosts and early fall frosts that damaged the crop before maturity. Consequently, shorter-season barley and rye became critical crops in the Holy Roman Empire. Barley was important as the mainstay in beer brewing. Rye had likely been unintentionally introduced as a weed in wheat seeds, but rye proved to be able to produce a crop even under cold conditions allowing it to become a major part of grain production on cool mountainous lands and in northern Europe. Rye became a starch source in making bread in the northern areas of Europe. A new, fourth grain became of increasing importance in feudal Europe – oats. Oats was likely carried from the Eastern Mediterranean as a weed that infested the wheat and barley crops. The value of oats, particularly as an animal feed, was discovered in Feudal Europe. Oats is particularly nutritious among the cereal grains and was readily consumed by horses.

The first sites for cropping in the Holy Roman Empire would have been localized on lighter sandy and loamy soils deposited in the flood plains of rivers. Due to periodic flooding and limited water-holding capacity of these sandy soils, the native vegetation of the flood plains would not have included an abundance of large trees. Hence, clearing of existing vegetation to create fields for the crops would have been less difficult on these flood plains than in the surrounding forests. Even so, clearing of the brush and shrubs from these flood plains would still have been a challenging task. Once cleared, the sandy and silty soils of the flood plain could be tilled using the shallow ard-like plow that had been used for centuries. By 1050 only about 12 to 15% of the land in Europe was used to grow crops. In 1086, the Domesday Book prepared as an inventory of new English territory for William the Conqueror estimated one-third of the sown land produced crops that would be used in making beer. This supplied two liters of beer daily for every man, woman, and child.

As the population of Europe increased and the demand for grain increased, cropping land was expanded beyond the lighter soils of the flood plains to the forested lands on heavier clay soils. In the northwest in the region that would become The Netherlands, low-lying areas were converted to agriculture by the construction of canals and dikes. By the 12th and 13th centuries the forests were cut, and the amount of land being cultivated grew to triple and quadruple, respectively, the cropping land area in 1050. This dramatic increase in cultivated land area was possible because a plow much more substantial than the ard had been developed to work the heavier soils of the originally forested uplands. It is likely that Slavs who migrated into central Europe in the 6th century introduced the moldboard plow (Figure 13.2). This was one of the crucial developments during the Holy Roman Empire. The moldboard was very heavy and had to be supported on wheels as it was pulled through the fields. The wheels on the plow would carry its weight and help to decrease the energy required to pull the plow through the soil. An additional advantage of the wheeled plow was that it could be set to give a fairly uniform depth for plowing the soil. The plow was designed to position a horizontal cutting shear in the soil at a depth of 10 to 20 centimeters (4 to 8 inches), which would cut the roots of the existing vegetation. Behind the iron-tipped shear was the moldboard that turned the soil over and buried the weeds under the soil. In a single pass with the moldboard plow, the existing native plants would be suppressed both by cutting their roots and by burying the tops of these plants. Control of weeds and clearing the land for sowing of the crop were effective, although turning the soil would bring deeply buried weed seeds nearer the soil surface to sprout with the crop seeds.

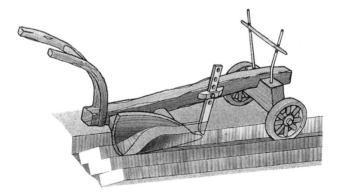

**Figure 13.2.** Wheeled plow used in Feudal Europe pulled by oxen hitched to the right end of the plow. (From Burke 1978, courtesy James Burke.)

The wheeled plow could not be pulled through the fields using human power as done with the ard on light soils. A team of two or four oxen would be used to pull the moldboard plow through the heavy soil of previously forested areas. A consequence of this dependence on oxen was that animal husbandry became important in central Europe. An additional consequence was that the plow and oxen team were not easily maneuvered, so it was desirable to plow fields laid out in long strips. Once the plow and oxen team were positioned, they would continue over a long distance in a straight line. Hence, fields were divided into the long, narrow strips which are still apparent in some parts of Europe.

Peasants had no voice in decisions about how the fields of the noble were to be farmed. Peasants were likely assigned to new fields each year depending on the crops to be grown and the work that was required. Consequently, there was little concern among the peasants for the fertility of the land on which the grain crop of the noble was grown. The focus of the peasants was more on small garden plots that they were commonly allocated to grow vegetables for their own consumption. Any collected animal manure would likely have been distributed on these gardens. Domestic animals were grazed in meadows and woodlands associated with each castle. Little manure was captured from the animals to apply to grain fields because the animals roamed free. Therefore, a major limitation in the production of grain crops during the Holy Roman Empire was a lack of nitrogen in the fields to support plant growth. It appears for much of this time the yields of these crops were below those that had been harvested in previous societies.

 None of the castle lands were fenced, so guards were needed to prevent freely grazing animals from straying into the crop fields. This became particularly important as the grains neared maturity and became attractive to animals. Consequently, peasants with dogs would have guarded the fields against encroachment by various animal pests. Even so, considerable crop losses to animals may have been a limiting factor in grain production during this era.

The original cropping system was one that had been used by the Romans, that is, a two-field rotation. Strips in a field would be either sown in a grain crop or left fallow. Therefore, only half the land was put in grain production each year. In the early years of the Holy Roman Empire, this would have been no problem since land was plentiful, although the forests would need to be cleared. If production needed to be increased, then more woodland would be cut to lay out another pair of fields for the standard rotation. Alternating between cropping and fallowing allowed nitrogen acquisition from rainfall and mineralization in the soil during the fallow period to provide some nitrogen for the subsequent crop. The downside of fallowing was that this practice gave an opportunity for the native plant species (weeds) to reoccupy the fields. The fields had to be plowed at least three times during the fallow season to suppress the reestablishment of weeds.

A dramatic upward spiral in total crop production and population occurred during this period of the Holy Roman Empire. Land clearing, of course, accounted for much of the increased production. Additionally, the innovation of a three-field rotation developed during this period also increased overall crop productivity in all but the most northern regions of Europe. Instead of only half the land being in a crop in one year as allowed by the two-field rotation, the three-field rotation allowed two-thirds of the land to be in production each year. The three-field rotation usually involved the growth of four grain crops in a repeating cycle (Figure 13.3). Field One would be sown in the early spring to a short-duration summer crop such as barley, oats, or a legume crop such as pea. These summer grains would have been harvested in late summer. Field Two, which had grown the summer grain in the previous year, would remain fallow until the winter grain of wheat or rye was sown in autumn. This fallow field would still have to be plowed several times during the summer to suppress weed growth. The final plowing would be done in the autumn before sowing of the winter wheat, barley, or rye. The final plowing would have been shallow to prepare a seed bed with furrows in which the hand-broadcast seed would settle. Field Three would be growing the winter grain sown in the previous autumn. The winter grain would likely be harvested in July and then the animals would be allowed to graze the straw in the field through the remainder of the summer. In the autumn, the field would be plowed for sowing of the summer crop in the following spring. Since the peak workload in each of the three fields occurred at different times of the year with the three-field rotation, a further advantage of this rotation scheme was that it spread out the work demands on the peasants.

| Field | Summer | Winter | Summer | Winter | Summer | Winter |
|-------|--------|--------|--------|--------|--------|--------|
| #1 | Barley | – | Fallow | Wheat | – | – |
| #2 | – | – | Barley | – | Fallow | Wheat |
| #3 | Fallow | Wheat | – | – | Barley | – |

**Figure 13.3.** Common three-field rotation scheme used in Feudal Europe.

At the beginning of the Holy Roman Empire, most people were peasants fully engaged throughout the year in the work of growing, harvesting, and handling grains for food. Less urgency in the work schedule existed during the winter in this cold climate for both the two-field and three-field rotations. Church observances of Advent and Lent during the winter were compatible with this period of lessened workload.

Oxen were a critical energy source for crop production throughout this Feudal Period. Until well into the Feudal Period, horses were used only for battle. The inability to use horses in the fields was a technological problem because available harnesses were not suitable for horses. The harnesses fitted around the necks of the horses cut into the ability of the horse to breathe when pulling a load. Not until the 10th century was a harness introduced, probably from China, distributing the pulling load on the shoulders of the horses so that it did not interfere with the animals' breathing. These new harnesses allowed the full strength and speed of horses to be used. In addition, iron horseshoes began to be used in the 11th century that protected the hooves of the horses. Large horse breeds developed specifically to carry fully armored knights into battle would have also had the strength for pulling heavy loads. However, horses were not used in plowing fields until late in this Feudal Period. The initial agricultural use of horses was to pull wagons to deliver grain to the emerging towns. The strength and speed of horses made them advantageous in hauling wagons in contrast to a large team of oxen. However, the transition from using oxen to using horses in plowing was very slow. While horses allowed for greater speed in plowing, they had to be fed supplemental grain because of their more-limited digestive capabilities and greater expenditure of energy. Oxen could be essentially fed by grazing in non-cropped areas throughout the year in much of Europe, but horses had to be supplemented with grain, usually oats.

Conversion to horse power for plowing was ultimately stimulated by an increasing demand for grain, even though there was little land left that could be easily converted to cropping. Also, the production system became more intensive with the adoption of the three-field rotation which required that tasks be done in a timely manner. Not surprisingly, the three-year crop rotation system was adopted concurrently with the increased commerce in grains. An important advantage of the three-field rotation was that it easily included the production of oats to support an expanding population of work horses.

Increased grain production seemed to only slowly impact the diet of the peasants. While the nobles and trade people ate wheat bread, the main food of the peasants remained gruel. A cauldron was heated over a fire, and each day new ingredients would be added to the mixture. The base of this gruel was wheat or other grains with meat possibly added on special occasions. Rabbits, chickens, and pigeons were raised to add – though only rarely – to the gruel mixture. Pieces of salted pork could also be added to the stockpot without much prior preparation. Apparently, the cauldron was rarely emptied; new flavors added to the pot would extend over several days even though the meat had already been consumed. The stockpot was always a source of a hot meal, even if it eventually was only a thick soup of grain.

Frumenty (or fermenty, fromity, or furmity) was also consumed by the peasants. Wheat grain was lightly crushed and soaked in water or milk near a fire in a shallow pan or bowl. Once the grain swelled and the mixture gelled, it could be eaten cold with fruit or honey, or warmed and mixed with the contents of the stockpot. Frumenty would have been a welcome addition to the usual meal taken from the cauldron.

Rye became an important grain for food in the more northern latitudes of Europe where wheat grew poorly. Wheat and rye were likely grown as a mixed crop in many regions simply because of the difficulty of maintaining a monoculture. This mixture of grain yielded flour that allowed the baking of rye bread: wheat provided gluten, and rye added starch. Rye bread became a staple food of northern Europe. In the most northern regions where wheat grew poorly, rye could still produce a plentiful crop. In these regions, rye flour was used to make a flat bread or cracker, which is now associated with Scandinavian cuisine.

During this period, pretzels appeared as a new style of bread. The history of pretzels is somewhat ambiguous but one story is that the pretzel originated in a monastery as a use of dough left over from making unleavened communion bread. Since pretzels could be made using only flour and water, they could be eaten during Lent at a time when it was forbidden to eat eggs, lard, or dairy products like milk and butter. One suggestion for the shape of the pretzel is that it represents arms crossed across the chest in prayer.

Beer, of course, was a central element of the daily diet of peasants. Indeed, Charlemagne recognized the importance of beer and decreed that planning for new towns must include breweries. Beer was fermented from all the grains produced in central Europe: barley, wheat, rye, and oats. Barley, because it was hardy in cold climates, was generally part of the mixture to provide the amylases to facilitate the breakdown of starch to sugars. Various mixtures of the four grains – plus a range of various flavorings including many fruits – resulted in a wide range of tastes. Brewing by the peasants was rudimentary. The mixture of grain meal and flavoring was warmed by the fire overnight to allow starch breakdown. After the mixture settled the following day, the liquid would be poured off and allowed to stand for a few more days to allow for fermentation. Each household produced this low-alcohol content beer, or "small beer".

A major advance in the brewing of beer during this period was made by those producing beer in large quantities for the inhabitants of villages and cities. Large monasteries were constructed across Europe, and one of their main sources of income was brewing and selling beer. Early in the Holy Roman Empire it was discovered that adding dried blossoms of hops

(*Humulus lupulus*) to the brew mixture helped to preserve the beer. Hops were added to the hot mash before yeast became active. The hops released two resins that inhibited growth of bacteria but were harmless to humans. Indeed, hops had long been consumed for their medicinal qualities including antiseptic, diuretic, and even supposedly an aphrodisiac. In making the first true beer, the oils of hops masked the sweet taste of the fermented grain. The somewhat bitter, sharp taste was appealing to many consumers and became a staple of brewing all across the Holy Roman Empire. The beer being brewed at this time would be the first in history recognized by modern drinkers as beer. Hops grew wild in what is now Eastern Europe on the trees of the forest around the river flood plains. A major industry developed to produce domesticated hops as early as the middle of the 9th century in Germany and eventually elsewhere across Europe.

Increasing grain output from the castle fields gradually freed small numbers of the population from field work. They were able to specialize in specific trades needed to grow crops and produce food for the castle community. The wheels of the heavy plows were constructed from wood by wheelwrights. Iron was worked by blacksmiths to make the iron shear tips for the plow. The processing of grain into beer and bread for the castle became specialized trades done by millers, brewers, and bakers in each castle community. In fact, family surnames originated from these professions, which in England were Smith, Miller, Brewer, and Baker, as Tom the brewer became Tom Brewer, for example. Towns, and eventually small cities, developed around the more productive castles. As the wealth and power of the noble's family and the clergy grew, additional trades developed including artisans who created the beginning of the Renaissance Period which emerged with the recovery from the Black Plague.

One of the important industries that developed during the era before the Black Plague was local brewing in the cities. Due to the difficulty in the logistics of transporting quantities of beer from the countryside to cities in a timely manner, local breweries emerged that purchased grain to brew beer in quantities to meet the immediate demand in the cities. Hence, grain was transported in quantity to the cities where beer was brewed to local tastes. Due to increasing demand for grain in cities for beer and for baking of bread, increasing trade in grain developed during the later period before the Black Plague. One of the important reasons for the increasing horse population and the need for growing oats was that horses were needed to pull the grain carts to the villages and cities.

The increasing trade in grain was likely a major contributor to the rapid devastation of the human population across Europe beginning in 1347. Carts hauling grain from the countryside to various cities also transported rats and their fleas; flea bites transmitted the bubonic plague to humans. This transportation network and the unwanted passengers rapidly spread the plague across Europe with devastating consequences. In some cases, whole villages were wiped out. It has been estimated that about one-third of the population of Europe died in a period of only five years from the bubonic plague.

The Black Plague was a major setback in increasing crop production and developing trades to make bread and beer. Depopulation of towns and cities meant that the market for grains disappeared. While the number of people to grow crops had also decreased, grain production exceeded the demand for grain. As a result, a major economic depression developed in Europe resulting in a steady decline in the price of grain over the next 150 years (1350 to 1500). Fields were abandoned and the impetus to expand production was gone. This era marked the true "dark ages" for much of Europe. Not until about 1550 with the repopulation of Europe and the rise of the Spanish and Dutch empires was there a stimulus to again expand cropping and agricultural productivity in Europe. The technology developed prior to the Black Plague was used in grain production to meet the grain demands of the expanding population of the cities of Europe.

**Sources**

Burke J (1978) *Connections*. Little, Brown & Company, Boston.

Dale T, VG Carter (1955) *Topsoil and Civilization*. University of Oklahoma Press, Norman, OK.

Evans LT (1998) *Feeding the Ten Billion. Plants and Population Growth.* Cambridge University Press, Cambridge, UK

Fussell GE (1976) *Farms, Farmers and Society: Systems of Food Production and Population Numbers.* Coronado Press, Lawrence, KS.

Kiple KF, KC Ornelas (2000) *The Cambridge World History in Food. Volume One.* Cambridge University Press, Cambridge, UK.

Larsen E (1977) *Food: Past, Present and Future.* Frederick Muller Limited, London, UK.

Montanari M (1994) *The Culture of Food.* Translated by C Ipsen. Wiley-Blackwell, Oxford, UK.

Slicher van Bath BH (1963) *The Agrarian History of Western Europe A.D. 500–1850.* Translated by O Ordish. St Martin's Press, NY.

Smitha F (n.d.) Map: Charlemagne's Empire, CE 800. [Online] Available at: http://www.fsmitha.com/h3/map04chrls.htm [Accessed 30 January 2010].

Tannahill, R (1973) *Food in History.* Penguin Group, UK.

# 14

# British

# 1700 to 1850

*The hand-mill gives you society with the feudal lord; the steam-mill society with the industrial capitalist.*

Karl Marx
*Poverty of Philosophy*

Population recovery from the Black Plague that decimated Europe in the middle of the 14th century was slow. Food production continued in the feudal system centered on castles and manors. The vast majority of people were still locked into an agricultural system that had changed little for hundreds of years. Peasants were under full obligation to the nobles and lords, crops were grown in three-field rotations, and life was sustained on a diet centered on porridge and beer. The long-standing importance of grain and beer to the British is documented by Clause 35 of the Magna Carta that King John was forced to sign in 1215. Although the Magna Carta is remembered as a critical document leading to democratic government, Clause 35 guaranteed that the same measure be used throughout the realm for grain and beer. In 1266 the basic troy measure was defined by law as a penny-weight sterling equivalent to the weight of 32 average grains from the middle of a ripe wheat ear; 7680 grains equaled one pound. These weights remained law until imperial measures were introduced in 1825.

Around 1700, however, new ideas for producing food began the movement to a Green Revolution. In England, in particular, new cropping systems were used that came to double grain yields and eventually freed nearly half the people from producing food. The serfs leaving the manors became factory workers of the Industrial Revolution. The combined agricultural and industrial prowess of Great Britain fed its emergence as the world's first global "superpower".

As with most revolutions in history, there were multiple causes for the agricultural revolution that occurred in England beginning about 1700. Political and economic pressure played a role because of the need for Great Britain to become self-sufficient in grain production. There was a nearly continuous state of war on continental Europe during the 18th century until the Napoleonic Wars ended with the defeat of Napoleon at Waterloo in 1815. Fortunately for Great Britain, the battlefields were all on the European continent, so British farmlands remained untouched. However, these wars meant that little or no grain could be imported from the continent. There was strong demand for domestic agricultural production in Great Britain.

To meet the demand for increased grain production, one of the first changes was to alter the historical crop rotations used in farming the land. The three-field rotation, in which there was a fallow year, was replaced by a four-field rotation. The highly productive four-field rotation was actually developed by Dutch and Flemish farmers in the Lowlands of Europe. A major motivation of the four-field rotation in the Lowlands was that it substantially increased the production of fodder for animals. Cows producing dairy products and meat for the expanding cities were the focus of agriculture in this region. The addition of a pasture cycle in the rotation that included a legume, i.e. clover, was important to achieving high productivity.

Importation to England of the four-field rotation concept resulted in the cropping sequence of wheat, barley or oats, turnips, barley or oats, and clover/ryegrass (Fig. 14.1). Wheat for bread and barley for beer were sold in the cities to provide the grain for the basic diet of the urban population. Turnips and oats were used as a winter feed for animals, and the pasture grasses were used for grazing. The addition of the pasture of clover/ryegrass, turnips, and oats allowed much higher year-round animal numbers. As a result, large numbers of animals need not be slaughtered in fall; there was now feed to carry them through the winter. Animals could be slaughtered year round, allowing a more frequent supply of meat for human consumption. Even more importantly, the animal feed supported large numbers of horses which provided the muscle to more efficiently farm the land. Not only did the two summers of clover/ryegrass in a field offer grazing for animals, but it also enriched the soil. Clover can fix atmospheric nitrogen into plant tissue, so when the pasture was plowed for the wheat crop the soil would be enriched with the nitrogen contained in the remaining parts of the clover and ryegrass plants. Additionally, concentrating the grazing animals on the pasture resulted in a concentration of manure on this field, putting nitrogen on the soil for use by the succeeding grain crops.

In England the four-field rotation was first established in the Norfolk region, so it was called the Norfolk system. The rotation sequence required land to be divided roughly equally among four fields to accommodate the two years of the clover/ryegrass pasture. The pasture cycle increased the fertility of the soil and resulted in large increases in yield. Grain yields increased by more than 50% in the 18$^{th}$ century climbing from the centuries-old European norm of one tonne per hectare to 1.5 tonne per hectare and greater. [A tonne is defined as the metric unit of 1000 kilograms. Tonne contrasts with ton which is the unit for 2000 pounds. One tonne equals 1.1 US ton.]

Implements were developed to facilitate the more intensive production of the four-year rotation. In particular, a sturdier plow was needed to handle the rigors of breaking up the sod left by the clover/ryegrass pasture. By the 1780's a plow made completely of iron was developed and manufactured on a large scale. A number of passes through the field in late summer was needed to completely destroy the pasture and prepare the land for the fall sowing of the small grain. Instead of oxen, the plows were pulled by a pair of horses and guided by a single ploughman. Horses could pull the plow faster than teams of four or six oxen. Further, a two-horse team was more easily maneuvered by a single person rather than the two people often needed for oxen teams. Fields could be much more rectangular, and the number of people involved in the demanding job of plowing the field was dramatically decreased.

| Field | Summer #1 | Winter #1 | Summer #2 | Winter #2 |
|---|---|---|---|---|
| #1 | Clover/ Ryegrass | Clover/ Ryegrass | Clover/ Ryegrass | Wheat |
| #2 | Turnips | Barley /Oats | Clover/ Ryegrass | Clover/ Ryegrass |
| #3 | – | Barley /Oats | Turnips | Barley /Oats |
| #4 | Clover/ Ryegrass | Wheat | – | Barley /Oats |

| Field | Summer #3 | Winter #3 | Summer #4 | Winter #4 |
|---|---|---|---|---|
| #1 | – | Barley /Oats | Turnips | Barley /Oats |
| #2 | Clover/ Ryegrass | Wheat | – | Barley /Oats |
| #3 | Clover/ Ryegrass | Clover/ Ryegrass | Clover/ Ryegrass | Wheat |
| #4 | Turnips | Barley /Oats | Clover/ Ryegrass | Clover/ Ryegrass |

**Figure 14.1.** The Norfolk four-field rotation allowing the growth of five crops per year.

Removing weeds from the crops was facilitated by sowing crops in rows, an innovation that had essentially been abandoned since the time of the Sumerians (Chapter 7). Horse-pulled seed drills were developed that dropped individual seeds in a row as it was drawn across the field. The spacing of individual seeds by the seed drills had the additional benefit of decreasing the number of seeds sown in comparison to hand broadcasting of seeds. Importantly, hoes pulled by a horse could be pulled through the space between rows to remove weeds. The width between rows, which persisted until the latter half of the 20$^{th}$ century, was fixed so that horses could easily walk between the rows. The labor-intensive activity of weeding was also dramatically decreased for the first time in history by using horse-pulled hoes.

Intensive cropping using the four-field rotation was, however, not compatible with the open-field layout that was the norm on castle and manor lands for centuries. Only animals belonging to the manor were to be grazed on the clover/ryegrass fields, so fences were needed to keep other animals off the fields. Fences made it possible to reserve grazing land for the animals owned by the lord of the manor. Construction of fences around fields was authorized by Parliament, which was composed of the lords and nobles who owned the manors. Beginning in the early 18$^{th}$ century and culminating by about 1850, Enclosure Acts in Great Britain were passed by Parliament allowing the construction of fences throughout the countryside. Stone fences and dense hedgerows constructed with permission of the Enclosure Acts are a hallmark of the English countryside to this day.

Another consequence of the Enclosure Acts was a consolidation of land in the hands of a limited number of nobles and lords. Those people with land holdings too small to be broken into the four or five fields needed for the four-year rotation were essentially forced to sell their land to the large manors. Labor-saving techniques were introduced on the newly consolidated and fenced manors to increase efficiency and crop yields. The amount of land which was plowed and weeded using horse power expanded during this period, resulting in a dramatic decrease in the need for human muscle. Fewer people were needed on the manor and there was no open "common" land on which peasants could eke out a subsistence living. The alternative to life on the manor for the rural peasants was to leave the land and migrate to the cities to work in the new factories of the Industrial Revolution. For the first time in history, the number of people engaged in food production decreased dramatically. Peasants traded long days of arduous work growing food for long hours of monotonous labor in factories. The large numbers of people displaced from the manors as a result of the agricultural revolution helped to provide the labor force needed by the Industrial Revolution. One of the remarkable changes during this period was that by 1830 only about 25% of the British population was directly involved in producing food.

Large yields on the manors of Great Britain resulted in a plentiful supply of grain. Wheat production was so high during the 1700s that Great Britain became a net exporter of grain during the window from 1697 to 1792. Domestically, the high production of grain caused bread to be readily available at cheap prices in the cities. Jobs in the newly developing factories were attractive because a factory laborer with a steady income had money to buy cheap bread and beer available in the cities. Meat also was available due to the animal production on the new, productive pastures. Those who moved to the cities often ate better than the peasants who stayed in the country and were vulnerable to the vagaries of each year's crop yields for the manor on which they lived. Increased agricultural productivity in the 18th century provided both the food and the people needed for the Industrial Revolution.

At the beginning of the 18th century, beer remained the beverage that all people drank. Benjamin Franklin wrote when he was in London in 1724 that the printers with whom he was working each drank beer "every day a pint before breakfast, a pint at breakfast with his bread and cheese, a pint between breakfast and dinner, a pint at dinner, and a pint in the afternoon about six o'clock, and another when he had done his day's work." This is a total of three liters of beer every day! During this period, the distinction between beer and ale was clarified according to the early dictionary of the English language published by Samuel Johnson (1755). Ale was the product of fermented malt while beer was brewed from malt combined with hops. Beer was thought to taste better than ale, and consequently the demand caused it to be more expensive.

The consumption of beer by rural people remained a mainstay of their diet, and the after-effects of alcohol on the brain were never of any particular concern, as had been the case since early civilization. Beer/ale was the safest, most nutritious, and surely most satisfying beverage for people to consume. However, to work in the factories a previously unknown requirement developed for mental alertness. While working with machines was usually very monotonous, the machinery demanded full attention. A brain that was dulled by alcohol would be unsafe for the worker and, more importantly to the factory owners, could result in slowed production, imperfect products, or damage to the machines. The laborers in the factory had to be weaned away from beer.

---

**Box 14.1.** Origin of Restaurants

The growing numbers of people in cities and the development of a middle class of artisans led to a whole new style in eating. Instead of eating all meals at home, this new middle class sought establishments that served a variety of food. Not surprisingly, the concept is attributed to a Parisian, Monsieur Boulanger, who set up his shop near the Louvre in 1765. He was legally constrained to serve only meat-based consommés that were not meals but medicinal supplements, i.e. "restaurants" intended to restore a person's strength. Later, Monsieur Boulanger boldly expanded his offering to include leg of lamb served in white sauce. This brought him into conflict with the catering guild for infringing on its monopoly to sell prepared meats. The subsequent lawsuit was settled in favor of Monsieur Boulanger – a very surprising decision in the age of dominant guilds. As a result, the monopoly of the catering guild was broken, and restaurants were allowed to sell a full range of sophisticated meals. Restaurateurs soon opened establishments all over Paris. Individual tables covered with cloths were introduced, as well as menus listing the multiple offerings.

---

What beverage would replace beer? Of course, water was still very dangerous as the pollution and sewage of the growing cities were dumped directly into streams and rivers. There were two key considerations in finding an alternative to beer consumption. One consideration was that the water in the beverage had to be made safe without involving alcohol. Boiling of the water was key to killing disease organisms in the water, although the existence of microbes had not yet been discovered. A second consideration was to provide a beverage with desirable mood-altering affects; acting as a stimulant for the brain, not a depressant. A hot drink made from boiled water and flavored with caffeine-containing substances was the answer.

By 1700, the stimulus of caffeine with the desirable flavor of cacao, coffee, and tea had been discovered by the upper class in Great Britain. Water was boiled, plant material was added to the water, and the drink was served hot. Cacao, coffee, and tea had to be imported so they were originally all very expensive. Through the 18$^{th}$ century these drinks, particularly tea, spread to the working classes laboring in the factories as the price of imported tea dropped. Even when diluted, tea imparted a satisfying flavor to the drink. In the cool and damp climate, hot tea became the favored drink throughout Great Britain.

Importation of tea into Great Britain came to be monopolized by the East India Trading Company. The East India Trading Company was controlled by the rich and powerful in Great Britain as the importation of tea increased from 50 tons in 1700 to 150,000 tons in 1800. A cornerstone of British foreign policy during the 18$^{th}$ century was the protection and exploitation of their monopoly on the tea trade, originally from China and then from India. A powerful naval force was built in part to protect the trade routes for tea. By the 19$^{th}$ century, Britain was the global superpower because of its dominance of the seas.

The tea trade monopoly was based on global exploitation. Tea was picked by hand from the tea bushes in the countryside of southern China and brought to Canton (Guangzhou) to sell to the British. The Chinese initially demanded payment in silver, and the price of silver in Europe rose dramatically. To avoid the rising cost of silver, the British began exchanging opium for tea. Opium was obtained in ports of what is now India and taken to China to exchange for tea. The large amount of opium traded in effect resulted in a huge portion of the Chinese population becoming addicted to the drug. Periodically the Chinese attempted to resist the British domination; the most notable conflict being the Opium Wars of 1839 to 1842. To avoid the continuing conflicts in China, the British eventually established tea production in India in the latter half of the 19$^{th}$

century. India became the jewel in the crown among the British colonies as it was the source of tea that consumers demanded in Britain.

Tea leaves boiled in water can result in a somewhat bitter drink, especially after the leaves have been in shipment from Asia over many months. An immediate solution to the bitterness of tea was the addition of large quantities of sugar. Also for factory workers, replacement of beer with tea meant a huge loss of caloric intake in their diet. Tea heavily laced with sugar helped to replace the calories formerly obtained from beer. Hence, the favored drink among the British became tea with large amounts of sugar. By 1801, on average each person consumed per year 1.1 kilogram (2.5 pounds) of tea and 7.7 kilograms (17 pounds) of sugar.

Another source of calories to replace those lost when beer was no longer consumed in quantity was bread covered with marmalade. Bread was cheap in the cities and was the main food for the factory laborers. To enhance the flavor of the bread, fruit marmalades were spread on slices of bread. Marmalade was made by adding large amounts of sugar to pieces of fruit. High concentrations of sugar, like large amounts of salt, create an environment in which it is difficult for bacteria to grow. The marmalades allowed fruits to be preserved (hence, they are sometimes called preserves) without refrigeration to be consumed at later dates. Orange marmalade made from fruit commonly obtained from Spain is still an important part of the British diet. By 1850, the usual meal of the British laborer was bread covered with marmalade and washed down with hot tea heavily sweetened with sugar.

The growing demand for sugar in tea and marmalade further expanded the trade routes of Great Britain. Sugar had been introduced by Columbus to islands in the Caribbean, and hence these islands became known as the Sugar Islands, providing the sugar consumed in Great Britain. However, the work of growing, harvesting, and processing sugarcane is among the most demanding of any crop. Further, the islands on which the crop was grown have a hot, humid climate that fosters a number of endemic diseases. Europeans would not take on the terrible work of growing sugarcane. Attempts to make slaves of the aboriginal people on the islands failed since these peoples were either killed by the diseases brought by Europeans or chose death rather than slavery. The alternative was to capture slaves in Africa and ship them to the Sugar Islands. Hence, the British established the terrible Trade Triangle; transporting slaves from Africa to the West Indies, shipping of sugar and rum to Great Britain, and carrying weapons and cloth back to Africa.

Sugarcane production required even more onerous labor than the production of grain crops. The soils in these islands were heavy clays. Trenches had to be formed in these clay soils into which stalks of seed cane were planted. Sugarcane grows slowly in its first months, so intensive efforts are required to prevent weeds from growing and overwhelming young sugarcane plants. Weeding was, of course, all done by hand. The hot, humid environment in sugarcane fields would be stifling, and the cane leaves are razor sharp. Sugarcane harvest required each stalk to be cut by hand using machetes and loaded on to carts. Sharp machetes led to accidents of deep wounds to arms and legs. Even more dangerous were jobs in the sugarcane mill where deaths were frequent from the cane crusher, boiling pots, and the heat itself. The work of growing and extracting sugar was brutal and done only by African slaves working under the threat of death. The life expectancy of slaves on the sugar plantations was short. Unlike on cotton plantations in later years, slaves did not make families and raise children. There was no hope for a child on a sugarcane plantation other than an early death. The owners of the plantation also saw no purpose in supporting children for many years when they would eventually live to work only a few years. There was a continuing demand for new slaves from Africa to work the sugarcane fields that supplied Europe.

An interesting historical aside was the adoption of the tea and sugar diet in the young American colonies. As in Britain, both tea and sugar were imported into the colonies to support their need for these essentials. Much of the tea and sugar that came into the American colonies was actually smuggled, thus avoiding British taxes. Not surprisingly, London saw smuggling as complete disrespect for British authority. In 1764, the British imposed the Sugar Act requiring governors to eliminate smuggling and collect the tax. While the British felt justified in collecting this tax to help pay the debt of the French and Indian War (1754-1763), which they viewed as an expense incurred in protecting the Americas, the colonists resented such domination from London. The Tea Act in 1773, along with several other colonial tax laws, further emphasized the dominion of England over the American colonies. The colonists strongly opposed these taxes as "taxation without representation", resulting in the Boston Tea Party in 1773 and eventually the American Revolution.

The outcome of the American Revolution also was very much linked to sugar and the sugar trade. After the victory of the British in the French and Indian War (1754-1763), the French ceded their claim to virtually all land east of the Mississippi River and Canada to the British. In exchange for this agreement, the French retained control of the islands of Guadeloupe and Martinique. These islands were much more critical to the French as sources of sugar, which had grown in demand in France, especially in elegant pastries. It seems, however, one of the reasons the French eventually offered support to the American rebellion was the possibility of strengthening their influence in the Caribbean. The arrival of the French navy at Yorktown in 1781 assured the surrender of the British forces and the ultimate establishment of the United States.

The period from 1700 to 1850 in Great Britain resulted in the most dramatic changes in agricultural technology up to that point in history, and the consequences were profound. New technologies resulted in a doubling of crop yields; rotations assured more nitrogen available for growing crops; and horse power allowed more land to be worked by fewer people. For the first time in history, food production was in the hands of a minority of the people, and laborers were freed to work in the factories of the Industrial Revolution. The British adapted their historic diet of beer and bread to tea and bread covered with marmalade. The momentum of the Industrial Revolution continued in Britain for approximately another 100 years beyond 1850. Domestic agriculture was no longer the foundation of British power; Great Britain shifted to a dependence on imports from its colonies to provide the wheat, tea, sugar, and other basic foodstuffs. Maintaining the naval power to assure shipping of exports from its factories and the import of foodstuffs was critical. Ultimately, diminishing naval power with the loss of ships in World Wars I and II weakened Great Britain, and its influence faded with the emergence of new industrial superpowers.

## Sources

Evans LT (1998) *Feeding the Ten Billion: Plants and Population Growth.* Cambridge University Press, Cambridge, UK.

Fussell GE (1976) *Farms, Farmers and Society: Systems of Food Production and Population Numbers.* Coronado Press, Lawrence, KS.

Hobhouse H (1987) *Seeds of Change: Five Plants that Transformed Mankind.* Harper & Row, NY.

Larsen E (1977) *Food: Past, Present and Future.* Frederick Muller Limited, London.

Ritchie CIA (1981) *Food in Civilization. How History has been Affected by Human Tastes.* Beaufort Books, Inc., NY.

Slicher van Bath BH (1963) *The Agrarian History of Western Europe A.D. 500-1850.* Translated by O Ordish. St Martin's Press, NY.

Tannahill R (1973) *Food in History.* Penguin Group, UK.

# 15

# Development of Science and Technology

# 1850 to 1950

*We've arranged a civilization in which most crucial elements profoundly depend on science and technology.*

Carl Sagan

*Any sufficiently advanced technology is indistinguishable from magic.*

Arthur C. Clarke

Crop yields in Great Britain by 1850 had risen to two tonnes per hectare, and these yields were produced with a drastic decrease in the number of farm laborers. Nevertheless, crop production was still based on human and animal muscle, and the diet of most people was still limited to a few choices. Scientific and technological developments in Europe and the

United States during the century from 1850 to 1950 provided the basis for a total revolution in food production, distribution, and consumption beginning in 1950. In particular, major advances were made in understanding the chemistry and biology associated with the critical requirements for crop productivity: nitrogen, water, and weed control. Additional technological developments had a far-reaching impact in changing completely the nature of food preparation, distribution, and consumption.

Experiments showing the critical need for nitrogen fertilizer for growing crops were begun before 1850. Jean Baptiste Boussingault, who was working in France, is credited with definitive early studies describing the nutrient composition of plants. In his 1844 publication *Traite d'Economie Rurale*, he reported that nitrogen was a major component of plants and that nitrate in the soil was critical for good plant nutrition. About the same time (1843), John Bennett Lawes established the Rothamsted Experiment Station on his manor in England to test in the field various soil fertility schemes to increase crop yield. He and chemist Henry Gilbert established long-term experiments testing a number of management practices including various crop rotations and nutrient fertilizer applications. These experiments provided scientific evidence about the various options for improving crop management. Similar long-term experiments were initiated by G. Morrow and M. Miles in 1876 at the University of Illinois, U.S. The Morrow Plots were designated a U.S. National Historic Landmark in 1968. Results from some of the surviving plots at Rothamsted and Illinois have been used to gain information about carbon storage in soils and possible impacts of climate change on crop performance. In 1888, the key role of bacteria in symbiosis with legumes to allow biological fixation of atmospheric nitrogen in plant organic compounds was described by H. Hellriegel and H. Wilfarth working in Germany.

New knowledge of the importance of nitrogen application to crop fields stimulated importation, particularly to Europe, of guano. Guano is nitrate-rich excrement from seabirds that had accumulated over the centuries on the dry, rocky shores of the Pacific coasts of Peru and Chile. Mined guano was shipped to Europe as a source of nitrate for both fertilizer and manufacturing of munitions. However, the long distances of transport around the southern tip of South America caused guano to be an expensive source of nitrogen fertilizer.

Since nitrogen was abundant in the atmosphere, an early research goal was to discover a chemical means to capture atmospheric nitrogen to provide an inorganic nitrogen fertilizer. At the beginning of the 20th century, Fritz Haber working in Germany showed that ammonia could be synthesized from gaseous nitrogen and hydrogen in the presence of an iron catalyst under very high temperature and pressure. Ammonia is the critical precursor to manufacturing nitrate for fertilizer and, of course, for explosives. Haber was eventually awarded the 1918 Nobel Prize for chemistry for his discovery of this reaction.

Commercialization of the reaction was made possible by the work of Carl Bosch for Badische Anilin und Sodafabrik (BASF) in Germany. Bosch solved the problems of dealing with a reaction at high temperature and pressure, and eventually replaced hydrogen in the reaction with natural gas, i.e. methane. A large ammonia production plant was opened in 1913 at Oppau, Germany, and became critical in the supply of nitrate for explosives for the Germans in World War I. One of the "prizes" sought by the Allies after World War I was technology from German factories used in producing nitrate. In 1931, Bosch received the Nobel Prize for chemistry. Military needs of World War II spurred construction of manufacturing plants worldwide for the production of munitions based on the Haber-Bosch process. After the war, these plants were used for large-scale production of nitrogen fertilizer.

The water requirement for growing plants was the subject of the first plant studies, and a scientific paper on the topic was presented to the British Royal Academy in 1699 by J. Woodward. A major impetus for studies on crop water use resulted from the western movement of agriculture in the U.S. to the High Plains and beyond. Some of the early studies were done at the Universities of Wisconsin, Utah, and Nebraska. One of the most comprehensive sets of studies was done by Lyman Briggs and Hiram Shantz working for the U.S. Department of Agriculture in Akron, Colorado. These early studies documented a very close relationship between growth and the amount of water required by crops. Increased crop growth necessarily required additional amounts of water.

In 1902 the first U.S. Reclamation Act began the process of damming waterways in western states to make water available for crop irrigation. Probably the most famous of these dams is the Hoover Dam authorized in 1928 to capture the water of the Colorado River. Water from these dams was supplied to fields by the ancient technique of gravity feed through a series of canals. Water flowed onto the fields, commonly in furrows between crop rows. The invention of the center pivot system in the late 1940s, credited to Frank Zyback in Nebraska, eventually resulted in a great expansion of land under irrigation. In the center pivot irrigation system water is pumped into a long line of sprinklers that might stretch up to one mile in length. The line of sprinklers rotates slowly around the center or pivot point resulting in a large circle of irrigated land. Today, the large green circles that result from this water application can be readily spotted when flying over these regions.

Weeds have been the bane of cropping since the beginnings of agriculture. Aside from the use of flooding by the Egyptians, the only method for weed control was physical removal of these unwanted plants. An entirely new approach to weed control emerged in the 1940s with the discovery of chemicals that selectively damaged individual species of plants. World War II gave a major stimulus to the development of such technology. Scientists working at the Rothamsted Experimental Station in the United Kingdom were charged with developing a chemical to control weeds and allow greater crop yields. They discovered 2,4-D (2,4-dichlorophenoxyacetic acid), which acts as a stimulus to meristem growth in broadleaf plants; the plants literally grow themselves to death. Since grass species are little affected, this herbicide had great utility in the management of major grain crops including maize, wheat, and rice. The mode of action is specific to plants, so the chemical itself has no impact on non-plant species. The chemical was commercialized in the U.S. by the Sherwin-Williams Paint Company in the late 1940s.

Another chemical that became widely used as a result of World War II was DDT (dichlorodiphenyltrichloroethane), which was used to control insects. The chemical was synthesized in 1874, but its insecticidal properties were not realized until 1939 by P.H. Muller working for the Swiss company Geigy. Geigy patented the chemical in 1940 and sent the information to its branches in Britain and the U.S. The chemical quickly became a central weapon in controlling mosquitoes spreading malaria and lice carrying typhus in both military and civilian populations. In 1945, DDT was released for civilian commercialization and quickly became the main method of insect control in agriculture. Dr. Muller was awarded the Nobel Prize in Medicine for his research on the use of DDT. Unfortunately, DDT proved to have serious negative effects on animal species in addition to insects.

The investigations described above relate directly to the age-old issues of nitrogen, water, and weed control in cropping. There were also other scientific and technological developments in the 1850 to 1950 period that had profound impacts on the surge in crop productivity and food supplies after 1950. Three areas of major interest are (1) plant breeding and genetics, (2) development of the internal combustion engine, and (3) food storage and marketing. Each of these is discussed briefly.

The Austrian monk Gregor Mendel published his studies on inheritance of various characteristics in pea plants (*Experiments in Plant Hybridization*) in 1866. Mendel was elevated to abbot in 1868, so his plant studies ended. It was not until 1900, when independently Hugo de Vries in the Netherlands and Carl Correns in Germany discovered Mendel's paper, that the relevance of his work to the emerging topic of genetics was recognized. Ironically, the significance of Mendel's work appeared to undermine Darwin's idea of natural selection, which was widely debated at the time. While theoretical genetics in the early 1900s was an area of intensive study, it was considered by many to have little relevance to the practical problem of breeding superior crops.

In the early 20th century, seeds for the next crop were simply selected from superior plants, not unlike what had been practiced since the beginning of agriculture. For maize, which was commonly grown in an open-pollinated environment, new genetic combinations were possible since pollen released into the air could be carried to the female reproductive organ on another plant. The mating was essentially random, but selection of superior plants at harvest did result in a slight tendency for improved maize lines. This approach at the beginning of the 1900s led to small yield increases in crops such as maize.

Hybrids were the next major development in crop improvement. Hybrids are a result of controlled mating between inbred parents. The inbred parents had to be developed by controlled self pollination, so only pollen from the same plant was used to fertilize its eggs. After a number of generations, both members of a gene pair would be genetically dominant or recessive. When two inbred parents are mated, i.e. a single cross, many genes would have at least one member of the pair expressing dominant characteristics, which are commonly beneficial and hopefully result in a superior hybrid plant. A common difficulty is that the seed yield for the female inbred parent is low due to the number of recessive traits being expressed in the mother plant. With poor seed yield in the production of hybrid seed, it was assumed that there was little commercial relevance in creating maize hybrids and the rediscovery of Mendel's work was thought to be of only academic interest.

Donald Jones, working at the Connecticut Agricultural Experiment Station in New Haven, suggested in 1918 a solution to the problem of low yield for hybrid maize seed. He proposed that commercial hybrids be developed from the mating of two single-cross parents. This scheme (Fig. 15.1) required four inbred parental lines to be developed as a result of approximately eight generations of self pollination. The four parents were selected to bring desirable traits to the subsequent matings. Two single-cross lines each derived from two inbred parents would be mated, resulting in a double-cross hybrid derived from four inbred parents. The single-cross female parent in the mating already would carry the capacity for substantial seed yield and thereby allow production of commercially viable amounts of seeds.

Still, the pollen transfer from the male parent to the female parent had to be carefully controlled in the production of inbreds, single crosses, and double crosses. The cost of controlled pollination in the field still limited the entry of hybrid maize into commercial markets. However, the brutal weather in the U.S. in the 1930's and the obvious yield superiority of double-cross hybrids resulted in increasing demand for hybrid maize seed. After about 1936, sowing of hybrid maize seed was widespread; sale of hybrid seed maize by both local and national plant-breeding companies became a new, economically viable agricultural industry in the U.S.

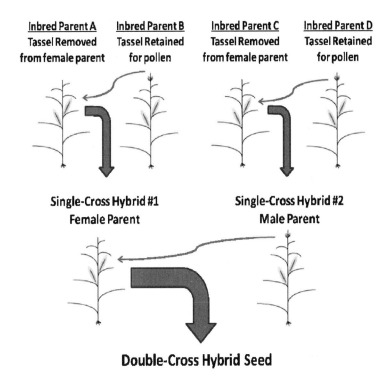

**Figure 15.1.** Schematic drawing of the production of double-cross maize seed from four inbred lines. The inbred parents are developed from eight generations or more of self pollination.

The second major technological development to be applied to crop production was the internal combustion engine, which replaced human and animal muscle in doing the work of growing crops. The internal combustion engine was developed in Germany in the late 1800s. Although Rudolf Diesel demonstrated his diesel engine fueled by peanut oil at the 1900 World's Fair in Paris, gasoline-fueled engines were used on the first tractors. In the U.S. several small companies developed tractors for farm use in the first two decades of the 20th century, their impact was minor. Only after the horse population was decimated in the fighting of World War I was there a boom in tractor sales. International Harvester (Farmall tractor) and Ford (Fordson tractor) became the dominant manufacturers in the 1920s. Launch of the Ford's Model A car in 1928 and the need for additional production lines for the car caused Ford to abandon tractor sales. Several other competitors entered the tractor business in the U.S. including Deere, Massey-Harris, and Case. The sales of tractors slowed during the Depression of the 1930s but resumed again to support agricultural production during World War II. By 1950, there were tractors on virtually all farms in the U.S., with a total of 3.5 million tractors.

The above technologies not only provided tools for large increases in grain productivity, but also could be applied to growing a whole range of vegetables and fruits. However, vegetables and fruits are often highly perishable and require careful preservation if they are to be consumed well after harvest. Food preservation had been a challenge through history. For much of history, the only means for preservation was drying fruits and salting and smoking meats. In the late 18th century, however, new technology began to emerge to enhance food preservation. Benjamin Thompson, Count Rumford, working for the Elector of Bavaria, invented the precursor of the bouillon cube made of solidified stock from trimmings of veal and pork. Thompson was an American, but as a British Loyalist he had to flee New England in 1776. In 1795 he took the position of aide-de-camp in Bavaria where he established workhouses for the poor and fed them his soup. Reconstituting the dehydrated stock in boiling water and adding a bit of grain resulted in an inexpensive soup providing a reasonably hearty meal.

In 1800, Napoleon Bonaparte – who had famously said that "an army marches on its stomach" – offered a prize of 12,000 francs for a method to preserve food to feed his army during his campaign to conquer Europe. Nicolas Appert developed the preservation technique in which food was put in wide-necked glass bottles, the bottles placed in boiling water, and then corked with wax while still hot. In 1810, Nicolas Appert was awarded Napoleon's prize, which he used to build a factory that burned down in

1814. A difficulty with Appert's method was that both the bottles and wax cork were vulnerable to cracking which would allow air to reach the food. Bryan Donkin, working in England in 1814, became the first to sell food in hand-formed tin cans which provided a more secure container. A small hole was left in the top of the tin can; after heating the can and its contents, solder would be used to seal the hole. These advances led to both home and commercial canning of foods.

Heating during canning can degrade taste and nutritional quality of food. It had long been recognized that cooling of foods maintained them for consumption over a long period. Root cellars dug in the soil were used to store foods, and ice collected during the winter and placed in the cellar would further extend the storage capability. Ice boxes were used in more affluent homes to store foods for short periods. One of the main commercial demands for ice was to keep beer chilled during storage and transport before it reached the consumer. Iced train cars were developed for transport of dairy products, meat, fruits, and vegetables long distances.

The truly significant advance in food preservation came, however, with the development of mechanical refrigeration. The basic process for chilling was understood to result from evaporation of liquids. Thermal energy is taken from the surroundings in the process of evaporation. Machinery was developed to take advantage of this process by allowing liquid to evaporate in tubes, cooling the tube and its surroundings. The evaporated gas would be re-condensed to liquid using external energy to complete the cycle. Early refrigerators used ammonia or sulfur dioxide as the coolant in the cycle, each of which could cause serious injury or death. Frigidaire, which was a part of General Motors at that time, discovered in 1928 that chlorofluorocarbons (CFCs) could be used in the refrigeration cycle and were safe with regard to direct human exposure. Large refrigeration units that maintained food at cool or freezing temperatures became a key component of food processing operations. Train cars to maintain either refrigerated or frozen foods were developed. Only much later was the destruction of the ozone layer caused by the release to the atmosphere of the CFCs used in these systems recognized.

In the 1930s household refrigerators began to be manufactured in the U.S., but major production occurred only after World War II. By 1950, more than 80% of the homes on farms had refrigerators and more than 90% of urban households had refrigerators. This allowed foods to be stored in the home for several days. It was no longer necessary to go to the market every day to purchase fresh food for the day. Household freezers also meant that meat and vegetables could be frozen and stored for long times. The ability to store foods for long periods introduced an entirely new approach to purchasing foods for the family.

Refrigeration allowed a whole new range of food products for the consumer to purchase and store in the home. Foods could be purchased with only one or two shopping trips a week, so food shopping was no longer a required daily activity. A response to this new style of food shopping was a consolidation of all foods into a one-stop market. In 1930, Michael Cullen opened a warehouse in Queens, New York, that has been described as the first "supermarket". Rather than individual foods offered in an array of specialty shops, all foods were brought together in a common store. The Great Atlantic and Pacific Tea Company, for example, began consolidating its many small A&P stores into supermarkets. The cheaper prices that could be offered in these high-volume supermarkets were especially appealing in the Depression years. By the 1950s supermarkets came to dominate the food sales industry. It has been suggested that an important contribution to the liberation of women was that they no longer had to devote large amounts of time to daily food shopping in a number of small shops.

## Sources

Davis, KS (1971) The deadly dust: The unhappy history of DDT. *American Heritage Magazine* 22(2).

Hager T (2008) *The Alchemy of Air.* Harmony Books, NY.

Krasner-Khait B (2000) The impact of refrigeration. *History Magazine*, March 2000. [Online] Available at: www.history-magazine.com/refrig.html [Accessed January 26 2010].

McCosh FWJ (1984) *Boussingault: Chemist and Agriculturist.* D. Reidel, Dordrecht, The Netherlands.

Rothamsted Research, UK (n.d.) *The Origins of Rothamsted Research.* [Online] http://www.rothamsted.bbsrc.ac.uk/corporate/Origins.html [Accessed 26 January 2010].

Sarkar P (2005) Scrambling for customers. *San Francisco Chronicle,* August 4. [Online] Available at: www.sfgate.com/cgi-bin/article.cgi?f=/c/a/2005/08/04/BUG7PE2DKH1.DTL [Accessed 26 January 2010].

United States Department of Interior, Bureau of Land Reclamation (2000) *Brief History of The Bureau of Reclamation.* [Online] Available at: http://www.usbr.gov/history/BRIEFHist.pdf [Accessed 26 January 2010].

United States Geological Survey (2000) *Irrigation Water Use.* [Online] Available at: http://ga.water.usgs.gov/edu/wuir.html [Accessed 26 January 2010].

White W (2008) *Economic History of Tractors in the United States.* EH.Net Encyclopedia, edited by Robert Whaples. March 26, 2008. [Online] Available at: http://eh.net/encyclopedia/article/white.tractors.history.us [Accessed 26 January 2010].

# 16

# Americans

# 1950 to present

*Agriculture allowed us to vastly multiply the populations of a few favored food species, and therefore in turn our own. And, most recently, industry has allowed us to reinvent the human food chain, from the synthetic fertility of the soil to the microwavable can of soup designed to fit into a car's cup holder.*

Michael Pollan
*The Omnivore's Dilemma*

Advances in science and technology since World War II have allowed an explosion in crop productivity and revolutionized foods in the U.S. Human and animal muscles were completely replaced by machines and chemicals. The number of people needed to grow food diminished to less than 2% of the population. The great majority of people was now isolated from the production of food, lived in urban areas, and obtained their food from supermarkets.

© T.R. Sinclair and C.J. Sinclair 2010. *Bread, Beer and the Seeds of Change: Agriculture's Imprint on World History* (Thomas R. Sinclair and Carol J. Sinclair)

The application of technology to crop and food systems in the U.S. was greatly stimulated by two factors. First was the burst in population labeled the Baby Boom which followed World War II. The second factor was a middle class growing in size and affluence, demanding more meat in their diets. Throughout history meat had been the food of the elite classes, and non-elite people only rarely had the chance to add meat to their porridge. Historically, with growing purchasing power one of the first life-style changes nearly universally made by humans is to increase meat consumption. In the U.S. the combination of increasing population and greater per capita demand for meat resulted in a huge increase in the demand for nearly all meats (Figure 16.1). Beef and chicken were the meats with the greatest increase in demand.

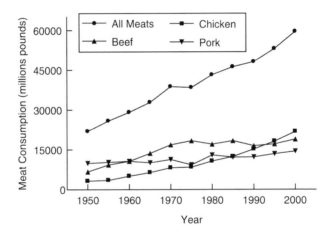

**Figure 16.1.** Total meat consumption per year in the U.S. since 1950. Consumption of beef, chicken, and pork is also shown individually.

To meet the increasing meat demand at a reasonable price, cattle could no longer be fed solely by grazing on pastures or chickens by scavenging in the farmyard. Animals were confined and fed with large quantities of grain, particularly maize and soybean. Hence, the demand for meat translated into a larger demand for grain production. Moving away from grain production mainly for direct human consumption, grains were needed to feed animals for meat production. The grain demand for meat production was huge; producing one kilogram of chicken requires two kilograms of grain as feed and one kilogram of beef produced in feedlots requires seven kilograms of grain.

By 1950 virtually all the land suitable for agriculture in the U.S. was under cultivation, so grain demand had to be met by increased yields. Significantly, crop yields in the U.S. in 1950 were not greatly different from those achieved in the United Kingdom in 1850. Emerging technology and scientific knowledge had to be put into practice to meet the increasing demand for food. A robust system in the U.S. of land-grant universities and state agricultural experiment stations was already in place to provide the intellectual resources for farmers. These institutions undertook the basic and applied research needed by farmers to increase yields. Further, an extension system was in place that included county farm agents working directly with farmers in applying technologies to local conditions. The research and extension structure working with farmers was able to keep U.S. farm production well ahead of consumer demand.

All technologies were applied in concert to achieve yield increases. A cornerstone for yield increase was an increased supply of nitrogen to crops, and nitrogen fertilizer applications increased steadily through the mid 1970s. As seen in Figure 16.2, the increase in maize yield up to the mid 1970s was associated with increased amounts of nitrogen fertilizer applied to the crop. More nitrogen was available to the plants to support greater growth and grain yield. Of course, plants genetically suited to take advantage of the increased availability of nitrogen and eventually use the nitrogen to grow more grain had to be developed in parallel with increased nitrogen availability. Plant breeding efforts developed new crop varieties that enhanced the use of the increased amounts of nitrogen fertilizer and were resistant or tolerant to various diseases and insects. In addition, it was no longer necessary to have crop rows sufficiently wide to allow a horse to pull a cultivator through the fields for weeding. Row widths were narrowed, and plants were bred to flourish with increased density to allow greater interception of solar energy and increased yields per land area.

Parallel technologies in fertilizer management and plant improvement led to the yield increases up to the mid 1970s (Figure 16.2).

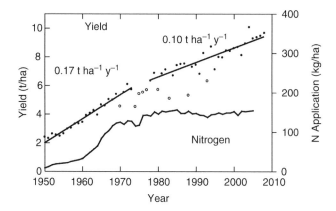

**Figure 16.2.** Maize yields per year shown on the left y-axis and nitrogen fertilizer application per year on the right y-axis in the U.S. since 1950.

However, increasing yields were not achieved without challenges. One challenge was meeting the need for large amounts of maize hybrid seed. New seed had to be generated each year to sustain the purity of the hybrid with the desired gene combinations. In the production of hybrid seed, rows of the female parent would be grown adjacent to rows of the male parent so pollen shed from the male parent would be readily received by the female parent, thereby resulting in a genetic cross. In the 1950s, the tassels on the female plant had to be removed before they produced pollen to prevent pollen from the female plant from reaching the silks to fertilize itself (selfing) (Fig. 15.1). Large numbers of young field workers were employed to remove tassels manually from all female parents from which the hybrid seeds were to be harvested. The production of double-cross hybrids was time consuming and costly. To decrease the labor requirement, inbred parents that gave especially high seed yields were identified. These inbred lines yielded enough seeds so that single-cross hybrids could be sold to farmers.

In the early 1950s, scientists at Texas A&M University discovered a gene that caused maize pollen to be sterile and this gene could be introduced into the female parent plant. Hence, with this gene in place it was no longer necessary to physically remove the tassels from the female parent. Not surprisingly, in the 1960s virtually all maize hybrids in the U.S. were soon produced using this genetic source of pollen-sterility.

Unfortunately, the genetic source of pollen-sterility allowed the plant to be susceptible to the fungus *Helminthosporium maydis*, eventually resulting in a major epidemic of southern corn leaf blight in the U.S. maize crop in 1970. While yields were dramatically decreased in that year, there were sufficient reserves of maize grain so that consumers were little impacted. Quickly, genetic sources of pollen sterility resistant to *H. maydis* were identified and used in producing new maize hybrids. However, this experience with southern corn leaf blight dramatically highlighted the danger of concentrating the genetic background of crops onto a small genetic base. An on-going effort is necessary to continually introduce new sources of genetic diversity into crops to maintain a broad genetic base and avoid vulnerability of the entire crop to a single disease or insect.

The ever-increasing applications of nitrogen fertilizer resulted in identifiable environmental consequences. Unfortunately, nitrogen in the soil from organic and inorganic sources is readily dissolved in water and, as a result, flows where water flows. Nitrogen can be lost from fields in both surface runoff from fields and in deep drainage of water under fields. Water and the dissolved nitrogen flow to streams and rivers. Eventually much of this water from the U.S. Corn Belt reaches the Gulf of Mexico via the Mississippi River. In the Gulf of Mexico, the nutrient-enriched water supports large algae blooms. When the algae die and sink, dissolved oxygen in the water is consumed in the decomposition of the dead algae. As a result the water becomes hypoxic and can no longer support much of the marine life. There is now an area of 15,000 to 17,500 square kilometers (6000 to 7000 square miles) at the mouth of the Mississippi

River in the Gulf of Mexico with substantially depleted marine life; it is referred to as a dead zone.

The loss of nitrogen from the fields also represents an important economic loss for farmers because fertilizer cost has become expensive with increases in the price of natural gas required to fuel the Haber-Bosch process. There have been important changes in the management of nitrogen fertilizer to increase nitrogen recovery by the crop and lessen nitrogen release to the environment. One major change is that the amount of nitrogen fertilizer applied to maize fields stabilized about 1980 at a national average of approximately 180 kilogram per hectare (160 pounds per acre) (Figure 16.2). The marginal return of nitrogen application above this level did not justify the expense to farmers of applying greater amounts. Not surprisingly, the rate of increase in maize yield per land area has dropped to less than 60% of the rate of increase prior to the mid 1970s; however, U.S. maize yields are still increasing, due to continuing improvements in the management of nitrogen fertilizer.

One of the improved management practices has been a shift in the timing of the application of nitrogen fertilizer for the crop. Earlier, most of the nitrogen was sometimes applied to fields in the fall or early spring prior to sowing the crop to spread use of machinery and time requirements in cropping. This resulted in a long residence time for the fertilizer in the field and increased the opportunity for nitrogen to move from the field. In many cases, the timing of fertilizer application has now been adjusted to better match the uptake capability of the crop. A small amount of nitrogen may be added to the soil at the time of sowing to support the plants in early growth. When crop growth is approaching its maximum rate about a month later, the bulk amount of nitrogen is applied to the soil to be available as the crop begins its rapid growth. New formulations of fertilizers are also being investigated to release nitrogen into the soil throughout the period of crop growth and to match even more closely the needs of the crop, avoiding high nitrogen concentrations in the soil at any one time.

Another nitrogen management technique that has been implemented in many locations is a decrease in the amount of tillage of the soil. At the beginning of this era (1950s), the moldboard plow used about 1000 years earlier in Europe was still the dominant mode of soil tillage. The moldboard plow buried weeds and residue from the previous crop so an exposed, uniform seed bed was prepared for the new crop. Plowing also exposed the dark soil to the incoming solar energy, increasing the rate of soil warming in the spring and allowing earlier sowing of the crop. However, a negative consequence of plowing was that turning the soil exposed it to oxygen, which stimulated mineralization of the organic matter in the soil to release nitrogen into the soil. Mineralization released nitrogen in the soil providing an additional boost in yields when nitrogen fertilizer application amounts were still low. However, the continuous mineralization and loss of organic matter were not sustainable; such a practice gradually depleted the productivity of the soil.

The energy cost of pulling large plows through the fields with tractors also became a major expense. An alternate low-tillage approach is now being used in which seeds are sown directly into the soil without plowing. A slit is cut in the otherwise undisturbed soil, and seeds are sown in the slit (Figure 16.3). Residue from the previous crop remains on the soil surface and decreases water evaporation.

**Figure 16.3.** Twelve-row no-tillage seed planter. The colters in front first cut a slit in the soil. The slit is opened by discs for dropping seeds from the seed box at the rear of the implement. Large rubber tires allowed the entire implement to be lifted to facilitate maneuverability. (Courtesy of Great Plains Mfg., Inc., Salina, Kansas, U.S.)

A major challenge in low-tillage management is dealing with weeds. By plowing with the moldboard plow, existing weeds would be killed and buried; the weeds in the low-tillage system need to be removed in a different way. The development of herbicides provided a key tool in dealing with weeds. There are several chemical options to kill the standing weeds at sowing, but these do little to suppress the emergence of a whole new population of weed plants after sowing the crop. Of course 2,4-D was used early in this modern era, but its use was difficult because it too could hinder the growth of young maize seedlings if applied at the wrong time. Delayed application of 2,4-D could allow growth of a weed population damaging to the young crop. Also, 2,4-D is much less effective against large weed plants. A family of triazine herbicides was eventually developed that could be effectively used with maize. However, triazines required careful management to be effective in a low-tillage system.

Applying low-tillage management to soybean, which is commonly rotated with maize in the U.S., was an even greater challenge. Herbicides were developed that could be applied to soybean to kill weeds commonly infesting these fields. However, these herbicides are expensive and require very careful, intensive management to be effective. A rain at the wrong time could allow weeds to escape herbicide damage, and soybean yield would be greatly depressed. Effective, widespread use of reduced tillage management did not become accepted until a new herbicide system based on glyphosate was developed.

Glyphosate, marketed by Monsanto Corporation under the name Roundup, proved to be a very useful all-purpose herbicide. Its mode of action is on a plant biochemical pathway synthesizing three key amino acids. Without synthesis of these amino acids the plants eventually die. Since the herbicide is selective for a pathway found in plants but not in animals, glyphosate has no influence on animals. The lifetime of the herbicide, particularly once it reaches the soil, is short (usually only a few days) so there is little concern about carry-over damaging a subsequent crop. The breakthrough in the use of glyphosate was the discovery by Monsanto of a gene that conferred plant tolerance to the herbicide. Biotechnology was used to insert this gene into high-yielding soybean varieties that were then resistant to glyphosate – marketed as Roundup-

Ready. The Roundup-Ready soybean field could be sprayed at any time to kill weeds and leave the crop undamaged. The ease in using this technology package greatly simplified low-tillage management, and it was widely adopted by farmers. The glyphosate-resistant gene has now been inserted into maize and cotton varieties.

Of course the Roundup-Ready approach, as with all biological solutions to controlling pests, is vulnerable to biological evolution that is unceasing in the field. Some existing weed species already had some tolerance of glyphosate. Not surprisingly, in response to heavy use of glyphosate weed mutants are also beginning to emerge on which glyphosate has limited or no effectiveness. The reaction has been to develop more complex herbicide mixtures and introduce genes into soybean that maintain its resistance to these mixtures. There is no final solution to the "biological wars" in crop fields, and newer approaches and technology will always be needed to maintain an advantage over crop pests.

Herbicides are formulated to attack targeted biochemical pathways found in green plants. Generally, the careful use of herbicides following labeled instructions has resulted in little damage to the environment or humans. One exception may have been Agent Orange, which was developed to destroy rain forests in the Vietnam War. Agent Orange was a mixture of 2,4-D and another general herbicide, 2,4,5-trichlorophenoxyacetic acid (2,4,5-T). While this second herbicide itself is relatively safe, its manufacture to include in Agent Orange may have left trace amounts of a dioxin in the mixture. Dioxin is very toxic to humans, and hence Agent Orange is believed to have caused human disease.

In contrast to herbicides, insecticides are often targeted to biochemical pathways that are shared by insects and other animal organisms. Hence, considerable caution is needed in the use of insecticides. As discussed earlier, the great success of DDT in the control of serious insect pests associated with disease resulted in its widespread use. One of the primary uses in agriculture was killing boll weevils in cotton. Cotton was on the verge of being lost as a crop because of the decimation caused by the boll weevil. Once DDT was available, it was sprayed on cotton fields repeatedly once cotton bolls (i.e. the fruit of cotton) appeared. While spraying of DDT was effective in suppressing the boll weevil, its release into the environment resulted in serious damage to other species. The problems of DDT in the environment, including its deleterious impact on bird eggs, were forcefully brought to the public's attention in 1962 by Rachel Carson in her book *Silent Spring*. The recognition of the potentially great harm from insecticides, in particular, resulted in major government regulation in the use of pesticides.

As with weed control, biotechnology has been employed to deal with insects. In this case, Monsanto Corporation isolated a gene from the bacteria *Bacillus thuringiensis* (Bt), a common soil bacterium that produces crystals containing proteins toxic to certain insects. The gene was inserted into cotton and dramatically decreased the number of sprayings required to control insects. Instead of 12 or more sprayings a season, Bt-cotton generally needed only three or four sprayings. This reduction achieved a cost savings to farmers with a much less intrusive impact on the environment. The Bt gene has now also been inserted into other crops such as maize as a potential insect control.

Application of nitrogen fertilizer and control of weeds and insects essentially eliminated some of the critical factors limiting crop yields through the ages. However, water availability remains the major factor now limiting crop yields in many regions of the U.S. Since much land on which grain crops are grown is not irrigated, year-to-year fluctuations in yield in the U.S. and elsewhere are closely tied to rainfall. The research described in Chapter 14 on crop water use demonstrated the physical link between growth and water loss, and there are no apparent options for breaking this relationship. Contrary to the arguments of some proponents of biotechnology, there is no biological solution to the physical link between growth and crop water loss. Crop yields will inevitably be limited by the amount of water available during the growing season for plant use. The solution to achieve major yield increases in the face of low water availability is irrigation.

The land area under irrigation surged in the U.S. after 1950, reaching a peak in about 1980. Much of the increase in irrigation depended on wells pumping water from underground reservoirs. The Ogallala Aquifer under the western High Plains in the U.S. was tapped to provide water for center-

Box 16.1. "Organic" Cereal Grain Production

Organic production is not a mainstream alternative for growing cereal crops for several reasons. The requirements for obtaining the U.S. Department of Agriculture certification label of Organic make widespread growing of cereal grains virtually impossible. Many of these restrictions are based on the idea of limiting the use of "artificial" approaches in growing the crops. The regulations are a curious set of dos-and-don'ts; for example, plant varieties generated from the transformation of a single gene are strictly prohibited, but plants resulting from breeding that alters hundreds of genes in unknown ways are acceptable. Certainly, all plants used in organic farming have already been genetically manipulated in their history. The regulations also include a long list of both synthetic and non-synthetic chemicals that can and cannot be used in organic farming.

The greatest challenge in producing organic cereal grains for human food and animal feed is satisfying the crop's basic requirement for nitrogen. Nitrogen has been the critical resource limiting crop yields throughout history. Plants must accumulate nitrate from the soil to form the proteins and nucleic acids that are essential for all life. In so-called non-organic farming, nitrate is produced from atmospheric nitrogen using the Haber-Bosch process (Chapter 15). In organic farming, nitrate is derived by the breakdown of organic material in the soil. In both cases plant uptake of nitrogen is a result of nitrate transfer from the soil to the plant.

The challenge in cereal farming is that modern yield levels require very large amounts of nitrogen. For example, wheat grain for flour contains approximately 22 grams nitrogen per kilogram (22 kilograms per tonne). If the yield of the crop is 8 tonnes per hectare (119 bushels per acre), then 176 (= 22 x 8) kilograms nitrogen per hectare is removed from the field. The removed nitrogen must be replaced in the soil for the next crop. In organic systems, the nitrogen can come from the application of manure, as has been practiced at least since the time of the Han Dynasty (Chapter 9), and/or from residue of a previous legume crop as done with the introduction of

**Box 16.1**. "Organic" Cereal Grain Production *continued*

the Norfolk system in Great Britain (Chapter 14). The limitation of these approaches is that they do not usually supply the large amounts of nitrogen needed for modern yield expectations. Application of manure and crop rotations including legumes allowed a doubling of yields from one to two tonnes per hectare in Great Britain in the 18th century. To increase yields from two tonnes per hectare to eight tonnes per hectare requires six times more nitrogen than the British provided to their wheat crops at that time.

As it turns out, manure and compost are not particularly good sources of nitrogen for grain crops. Nitrogen concentration of manure from swine and poultry is usually less than 1%; manure from ruminant animals such as cattle provides even less than 0.5%. Therefore, replacing only the 176 kilograms per hectare of nitrogen removed in the grain harvest requires 20 tonnes or more of manure per hectare applied to fields. This estimate does not account for the fact that much of the nitrogen in the manure will not be available to the crop. Unlike manufactured nitrate which can be applied as it is needed by the crop, nitrate from manure is slowly released year round. As a result, nitrate moves from fields year round, contributing to increased nitrate in waterways.

The bottom line is that only limited production of organic cereals is possible, and the cost will necessarily be elevated over crops fertilized with manufactured nitrate. For most cereal production, there are simply not sufficient organic sources to adequately fertilize crops. The organic approach to fertilization is more appropriate for fruits and vegetables in which the extent of land area used in growing these crops is much smaller than for cereals, and the annual nitrogen requirements are much less. However, for all organic fruits and vegetables it is necessary to confirm that appropriate practices have been used by the grower in handling the manure and compost for fertilizing the crops so human-disease organisms have not infected the produce. Bacterial contamination can present a health hazard from consumption of organic products that are not properly sanitized.

pivot irrigation from South Dakota to New Mexico. However, the water in this aquifer was "fossil water" not readily recharged with new water. Not surprisingly, heavy pumping of water caused the water levels in the aquifer to decrease steadily. In many areas, pumping of the Ogallala Aquifer has now been abandoned because of the high cost of pumping water from great depths. Farmers in this area are seeking "dry land" cropping alternatives. The total amount of irrigated land in the U.S. has remained constant since about 1985 as decreases in irrigation in the High Plains have been offset by increased irrigation in more eastern areas of the U.S.

The combined application of all the new technologies resulted in large increases in crop yield. One indicator of the success in increasing agriculture productivity in the U.S. has been the consistent positive trade balance in agricultural commodities (Fig. 16.4). Export of agricultural commodities was small until the early 1970s, when the world began importing large amounts of U.S. maize and soybean. Since 1999 there has been a steady increase in the value of U.S. agricultural exports – with China becoming a major consumer. In the last few years exports to China have increased dramatically. Also, the import of agricultural products to

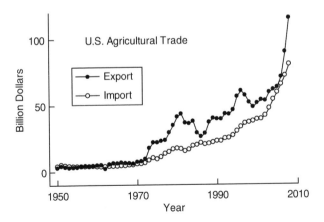

**Figure 16.4.** Annual U.S. export/import of all agricultural commodities since 1950.

the U.S. has risen steadily over this period as Americans enjoy a variety of foods produced around the world. The fact that agricultural exports have exceeded imports is very important to the U.S. balance of international trade.

The impact of the use of machinery powered by fossil fuels combined with the above technologies allowed a very small fraction of the U.S. population (less than 2%) to produce an abundance of foods. Modern supermarkets offer urban consumers abundant food of all descriptions at unprecedented low prices. Food used to be near the top of the list of expenses in the American household budget. Now, the fraction of the family's budget spent in the supermarket is near the bottom of the list. This small expenditure buys food produced across the U.S. and beyond. Consumers have become isolated from the vagaries of growing crops because there is usually a range of alternatives in production areas and food sources. In modern times, people do not go hungry because of a failure in the production of local crops. Hunger still exists in the U.S., but it is an "economic starvation" where an increasing number of people may not have enough money to purchase the riches of food available in the supermarket.

The abundance of fruits, vegetables, meat, and cereal products has resulted in remarkable improvements in people's health. The stature of humans has increased dramatically since the majority of children receive adequate protein and caloric intake. The nutrient-deficiency diseases discussed in Chapter 5 now occur only rarely in industrialized countries. Along with tremendous advances in medical care, the elimination of hunger and starvation has allowed life expectancy to soar.

However, the abundance of food has resulted in a whole new set of major health problems. The human body that evolved with a hunger for carbohydrates and fats to survive the physical demands of life and to be prepared for future famine is now in an environment where there are few physical demands, and for most people starvation never comes. Consumption of large amounts of kilocalories combined with sedentary life style has resulted in what is sometimes called "over nutrition"; that is, obesity has now become a major health concern in the U.S. Obesity is gauged by a person's body mass index, which is the ratio of the person's weight (kilograms) divided by their height squared (meters squared). Therefore, a person 1.8 m tall (5 feet, 9 inches) and weighing 70 kilograms (154 pounds) would have a body mass index of 21.6, which is in a healthy index. A person of the same height but weighing 95 kilograms (209 pounds) would have a body mass index of 29.3, which is on the border between overweight and obese. About one-third of American adults are now classified as being obese, i.e. a body mass index of greater than 30. The list of diseases as a result of obesity is long. Coronary disease, type 2 diabetes, and hypertension are diseases that are often directly linked to obesity. The risk of other diseases is increased with obesity, including cancers (particularly endometrial, breast, and colon), dyslipidemia (high cholesterol or triglycerides), liver and gallbladder disease, sleep apnea, osteoarthritis, and gynecological problems. An amazing fact is that in a few short years humans have gone from a major concern of obtaining enough kilocalories for survival to a near crisis in the U.S. from consuming too many kilocalories.

Inexpensive food itself has resulted in major changes in life style; people in the U.S. now spend much of their eating budget to have other people prepare their food. Nearly all products on the supermarket shelves are foods that have been processed to one degree or another to facilitate meal preparation. Pre-prepared foods range from cleaned and polished fruits and vegetables, boxes of macaroni and cheese ready for cooking, and microwave-ready meals. Even fully prepared meals are readily available in some supermarkets. The actual cost of the raw food ingredients in many supermarket products is now a minor portion of the total price.

Restaurants of all descriptions are a new industry having fully emerged in the era of cheap food. Fast food is now a staple of many U.S. families' eating habits. In 2000-01 a survey showed American men ate restaurant fast food about twice a week; white women ate somewhat less often at 1.3 times per week. The greatest consumer of fast food is the American child. A report released in 2004 showed that 30% of children in the 4- to 19-age bracket ate fast food *every day*. There are now 50% more people employed in the U.S. as fast-food cooks than the number of full-time farmers (612,000 vs. 393,000, *Time*, 26 Nov. 2007). In addition many families now expect to eat at a wide selection of family-style restaurants. Like supermarket products, the actual cost of the food ingredients is only a small element in determining the final cost of a restaurant meal. Therefore, American restaurants offer huge servings to attract customers, and diners have become accustomed to eating too-large portions of food. Consumption of these prepared meals adds to the challenge of matching caloric intake with physical activity.

Previous chapters in this book concluded with a description of changes in food production systems associated with the deterioration of the Golden Age of each society. From the crop and food perspective, there is no evidence that the U.S. is near the end of its Golden Age of power and influence. There have been challenges in producing grains at unprecedented levels, but these have been met. The dietary problems in U.S. society have resulted in large part as a result of the abundance of food. Efforts are being mobilized for people to better match their food intake to their activity level.

It is premature to speculate about the end of the Golden Age of the U.S. when we are only a little more than a half century into this era – in contrast to previous societies whose prominence lasted for centuries and in some cases more than a millennium. However, we are now living in an age where cropping, food supply, and environmental issues are global, not restricted within the borders of individual societies. The cost of foods is now set by events across the globe. Recent increases in food prices in the U.S. resulted from such distant events as increased demand for meat in China and drought in Australia. The era of U.S. global dominance may be changing to one requiring a planetary perspective: the role of the U.S. appears to be evolving to one of partnership with the countries of the world to assure food supply and environmental security. While it is impossible to predict the future, one desirable scenario could be a transitioning of the U.S. from a classical Golden Age to one of cooperative leadership to assure a Golden Age for the entire planet.

Whatever the future brings, modern agriculture can benefit from the lessons of history. While the illusion exists, as it did for past societies, that grain and food availability is in abundance, agricultural systems are inherently vulnerable to change in climate, soil degeneration, and attack by pests. Land resources must be protected. We cannot destroy the productivity of the land and simply move to new areas as the Bantu had done. Soil and water resources must be used wisely. We have no Nile River such as nurtured the ancient Egyptians by depositing a new layer of soil each year and fully filling the soil with water. Water resources must be managed wisely to avoid environmental degradation such as happened with the Sumerians in the salting of their soil. The Greeks and Romans ignored their soil resources until erosion forced them to become dependent on food imported from large distances. A strong knowledge base is required to deal with adverse environmental challenges such as flooding experienced by the Chinese. Research and development resources need to be in place to deal with unexpected challenges. Of course, feudal Europe had no knowledge to deal with the sudden appearance of the bubonic plague. On the other hand, the U.S. had the capability to react rapidly when Southern Corn Leaf Blight attacked the maize crop.

Climate change may well be another major challenge in future food production. The Maya suffered a loss of their Golden Age due to a sudden shift in rainfall patterns. Potential shifts in the amount and timing of rains could well be the major consequence of future global climate change. While changes in temperature and atmospheric carbon dioxide concentration are the focus of much discussion, for crop plants the projected changes in these two variables are likely to be accommodated by genetic or management modification to minimize negative impacts on crop productivity. On the other hand, a change of a few percent in the amount of rainfall or in the interval between rainstorms can have a major influence on crop yields. Unfortunately, there is no certainty about how rainfall patterns will be altered as a result of climate change. The recent report from the United Nations Intergovernmental Panel on Climate Change predicts a decreased amount of summer rainfall in the central region of the U.S. where the bulk of U.S. crops are grown. Aggressive research is needed to understand the response of crop plants to less rain and to explore options for adapting crops to a new climate.

Another important lesson of history is the need to maintain a reasonable balance in the dependence on local and imported resources. The Athenians and Romans became fully dependent on the importation of grain. Protection of crop production in distant lands and a dependence on long-distance shipping caused vulnerabilities for both societies, eventually undermining their power. The British also became dependent on far-flung lands for their food. First, tea and sugar were brought from distant places, and then basic grains were imported from Canada, Australia, and the U.S. British power was built in part on an ability to protect these shipping lanes. While the U.S. is well endowed to sustain food production within its borders, its large grain surpluses of the 1950s to 1990s are gone. Along with the loss of grain surpluses, the food production system has become totally dependent on machines. It is the fuel for these machines that is now imported over long distances and creates vulnerability not unlike the Athenians, Romans, and British experienced importing grain. Is it possible to drastically decrease the dependence of the food system on these imported fuels?   Can plant production capability be effectively and sustainably applied to producing biofuels?   In search for some of the answers to these key questions about the future we return to ages-old issues of growing and using crop plants.

**Sources**

Bowman SA, SL Gortmaker, CB Ebbeling, MA Pereira, DS Ludwig (2004) Effects of fast-food consumption on energy intake and diet quality among children in a national household survey. *Pediatrics* 113:112-188.

Carson R (1962) *Silent Spring.* Houghton Mifflin, NY.

Intergovernmental Panel on Climate Change, Working Group I (2007) *Chapter 11. Regional climate projections.* [Online] Available at: http://www.ipcc.ch/publications_and_data/publications_ipcc_fourth_a ssessment_report_wg1_report_the_physical_science_basis.htm [Accessed 26 January 2010].

Lal R, DC Reicosky, JD Hanson (2007) Evolution of the plow over 10,000 years and the rationale for no-tilling farming. *Soil & Tillage Research* 93:1-12.

Pereira MA, AI Kartahov, CB Ebbeling, L Van Horn, ML Slattery, DR Jacobs, Jr., DS Ludwig (2005) Fast-food habits, weight gain, and insulin resistance (the CARDIA study): 15-year prospective analysis. *Lancet* 365:36-42.

Pollan M (2006) *The Omnivore's Dilemma.* The Penguin Press, NY.

Rogers JS, JR Edwardson (1952) The utilization of cytoplasmic male-sterile inbreds in the production of corn hydrids. *Agronomy Journal* 44:8-13.

Sinclair TR, LC Purcell, SH Sneller (2004) Crop transformation and the challenge to increase yield potential. *Trends in Plant Science* 9:70-75.

Tatum LA (1971) The southern corn leaf blight epidemic. *Science* 171:1113-1116.

Williams GM, R Kroes, IC Munro (2000) Safety evaluation and risk assessment of the herbicide Roundup and its active ingredient, glyphosate, for humans. *Regulatory Toxicology and Pharmacology* 31:117-165.

# Epilogue

# Future of Grain Fermentation

Fermentation of grains to produce ethanol, i.e. beer, was a major factor in stimulating the beginnings of agriculture. Throughout history ethanol production played an integral role in sustaining most societies. Beer production was desired because it provided readily digestible calories and nutrients, contributed a comparatively safe beverage, and offered mood altering qualities. Grain fermentation was critical in "fueling" the human muscles that until recently were the basis for nearly all activity.

The per capita consumption of beer decreased with emergence of the Industrial Revolution. A wide range of safe food and beverages in modern, industrialized countries meant only a small fraction of grain production was used for fermentation. The bulk of grain production was shifted to feed animals in the production of meat. Now, in the last few years there has been another dramatic shift in the U.S. in the use of grains, particularly maize. Grain is again being grown to produce ethanol at levels unparalleled in history. The irony is that, instead of fueling human muscle, ethanol is now being used to provide the liquid energy required to fuel machines. The increase in the production of fuel ethanol was comparatively slow and steady up to about 2002, when just over eight billion liters (two billion U.S. gallons) were produced (Fig. E1). Then, in the period 2006 to 2008 there was a dramatic increase spurred by subsidies from the U.S. government to construct fermentation facilities. In 2008, U.S. fuel ethanol production reached about 37 billion liters (over nine billion U.S. gallons).

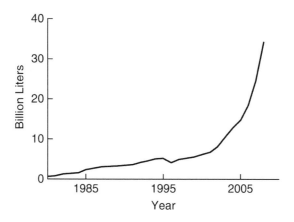

**Figure E.1.** Ethanol biofuel consumption per year in the U.S. since 1980.

While low oil prices in 2009 slowed construction of new ethanol fermentation facilities in the U.S., further large increases in ethanol fermentation are projected for the future. The 2007 U.S. Energy Independence and Security Act set the target for 2022 for national ethanol production at 144 billion liters (36 billion gallons) a year, nearly quadrupling current output. This level of ethanol production is equivalent to about 15% of the total current consumption of gasoline in the U.S. It is likely that most ethanol will, in fact, be consumed as a mixture of ethanol and gasoline. A mixture of 90% gasoline and 10% ethanol (E10), for example, works well in current engines. Higher ethanol concentrations than used in E10 are also possible for use in current gasoline engines. The use of ethanol has the major advantage of replacing in gasoline the additive methyl tertiary-butyl ether (MTBE), which is used as an octane booster. The problem with MTBE is that even very small quantities in water as a result of gasoline spills affect the taste of the water and make it non-potable.

In addition to setting the target for overall ethanol production for 2022, the U.S. Security Act specifies the source of the feedstock to be used in fermentation. No more than 40%, or 57.6 billion liters (14 billion gallons) ethanol per year, of the target is to be derived from grain sources. Maize has accounted for virtually all the past grain feedstock in ethanol fermentation. Assuming 400 liters (100 gallons) obtained by fermenting one metric tonne of maize grain, a total of 144 million tonnes (158 million English tons) of maize grain will be needed to meet the target ethanol production. At this level, nearly half the maize crop will be needed for ethanol production. The target for grain feedstock in the U.S. Security Act presents a major challenge in maize production to meet all demands including maize consumption as human food, animal feed, and export.

The remaining 86.4 billion liters (22 billion gallons) of ethanol according to the U.S. Security Act must be derived from non-grain feedstock. It is proposed that whole-plant harvests, consisting mainly of stems ranging from grass species to trees, are to be used as the main source of raw materials for fermentation. Since stems contain large amounts of cellulose and hemicellulose, digestion of the stems by chemical or enzymatic treatment to release sugars from these compounds would in principle provide the feedstock for fermentation. There are, however, several technological and economic challenges. Cellulose and hemicellulose in the stems are the main components of cell walls. To assure a strong cell-wall structure enabling plants to remain standing when subjected to strong winds and to withstand the negative pressures of water flow in stems the cellulose and hemicellulose are tightly interlaced with lignin and proteins. Unfortunately for ethanol production, lignin and proteins must be removed from the cell walls to allow digestion of the remaining cellulose and hemicellulose. Procedures currently available to accomplish this task are quite expensive. Once released from the cell walls, the strands of cellulose and hemicellulose must be unwound before they can be disassembled into their component sugars. These strands are tightly held together to give them strength, and the unwinding is much more demanding than for starch which can be opened up by simply soaking and heating. Finally, the sugar obtained from hemicellulose is unlike that from cellulose and starch in that it is a five-carbon sugar. Existing yeasts have little ability to ferment five-carbon sugars. While biotechnology is being applied to develop organisms that can break down the five-carbon sugars from hemicellulose, commercial methods for using hemicellulose have not yet been developed.

The goal of producing 86.4 billion liters (22 billion gallons) of ethanol per year from non-grain feedstock will result in one of the largest, most rapid changes in land use in history. The land area required to produce the non-grain feedstock is larger than that currently used for any single crop in the U.S. Land area estimates to produce the non-grain feedstock run as high as 48 million hectares (118 million acres), or more than 50% greater than any single crop currently in production. Even to produce the non-grain feedstock on this land area, intensive cropping practices will need to be applied. Economic analysis indicates that the minimum yield for non-grain feedstocks needs to be 9 tonnes of plant material per year per hectare (8 English tons per year per acre), a level of total production of plant material not achieved in history until the last 60 years. As discussed in Chapter 15, this higher plant production was only achieved with the application of nitrogen, and sometimes water, to crops. The basic needs for nitrogen and water to achieve high yield levels by these plants will exist even if grain is not the primary product.

Non-grain feedstock for ethanol fermentation will require management not unlike that required when growing grains. Nitrogen application will be needed on the lands producing non-grain feedstocks to reach the 9 tonnes per hectare yield. The application of nitrogen to a greatly expanded land area will further acerbate environmental and ecological concerns. More nitrogen will flow into waterways and aquifers, and increased amounts of oxides of nitrogen will be released into the atmosphere to act as greenhouse gases. Also, stimulated plant growth on the lands producing non-grain feedstocks for ethanol will require more water. This greater water demand means that lands in arid regions with low rainfall will simply not have sufficient water to reach the nine tonnes per hectare yield. Even in humid climates, stimulated plant growth necessarily results in greater evaporation of water from the plants. As a result, the hydrological balance in watersheds will be altered, decreasing water levels in streams, rivers, and aquifers. This use of water to obtain ethanol may conflict with other demands for fresh water.

Harvest each year of large amounts of plant material will result in additional challenges. While harvesting only grain allows the remainder of the plant material to stay in the field to conserve soil, harvest of most plant material can result in new challenges. There can be increased risks of soil erosion and long-term decreases in the amount of organic matter left on the land for incorporation into the soil. The sheer bulk of plant material for non-grain feedstock can be a problem in harvesting, storing, and transporting the plant material. Ethanol fermentation facilities will need to be situated near non-grain feedstock fields, making each facility vulnerable to local fluctuations in plant yield. In years of local drought, for instance, obtaining the bulky feedstock from distant locations could be difficult and expensive.

A number of essential technological challenges need to be overcome for both grain and non-grain ethanol production. Challenges in growing crops and also in using stems in ethanol fermentation must be resolved. For the moment, the fermentation of sugars and starch of grains remains closest to economic viability in the production of ethanol. Grains for bread and beer are now being diverted to the production of ethanol. The next chapter of the imprint of grains on history is now being written.

**Sources**

Sinclair TR (2009) Taking measure of biofuel limits. *American Scientist* 97:400-407.

U.S. Congressional Research Service. *Energy Independence and Security Act of 2007: A Summary of Major Provisions* (RL34294; Dec. 31, 2007), by Fred Sissine. Text in: LexisNexis® Congressional Research Digital Collection. [Online] Available at: http://energy.senate.gov/public/_files/RL342941.pdf [Accessed: 26 January 2010].